# Creating Effective Information Technology Solutions

*The mission of the Awwa Research Foundation (AwwaRF) is to advance the science of water to improve the quality of life. Funded primarily through annual subscription payments from over 1,000 utilities, consulting firms, and manufacturers in North America and abroad, AwwaRF sponsors research on all aspects of drinking water, including supply and resources, treatment, monitoring and analysis, distribution, management, and health effects.*

*From its headquarters in Denver, Colorado, the AwwaRF staff directs and supports the efforts of over 700 volunteers, who are the heart of the research program. These volunteers, serving on various boards and committees, use their expertise to select and monitor research studies to benefit the entire drinking water community.*

*Research findings are disseminated through a number of technology transfer activities, including research reports, conferences, videotape summaries, and periodicals.*

# Creating Effective Information Technology Solutions

Prepared by:
**Terry Brueck**, **Bruce McClung**, **Melanie Rettie**, and **Rich Lackmann**
EMA, Inc., 1970 Oakcrest Avenue, Suite 100, St. Paul, MN 55113-2624

Sponsored by:
**Awwa Research Foundation**
6666 West Quincy Avenue, Denver, CO 80235-3098

Published by:

Awwa Research Foundation
Advancing the Science of Water®

IWA Publishing

**DISCLAIMER**

This study was funded by the Awwa Research Foundation (AwwaRF). AwwaRF assumes no responsibility for the content of the research study reported in this publication or for the opinions or statements of fact expressed in the report. The mention of trade names for commercial products does not represent or imply the approval or endorsement of AwwaRF. This report is presented solely for informational purposes.

**Published by IWA Publishing, Alliance House, 12 Caxton Street, London SW1H 0QS, UK**

Telephone: +44 (0) 20 7654 5500; Fax: +44 (0) 20 7654 5555; Email: publications@iwap.co.uk;
Web: **www.iwapublishing.com**

AwwaRF report number 90977F. Originally published by AwwaRF for its subscribers in 2003.
IWA Publishing version published 2004.



ISBN 1 84339 877X

# CONTENTS

# TABLES

# FIGURES

# FOREWORD

The Awwa Research Foundation is a nonprofit corporation that is dedicated to the implementation of a research effort to help utilities respond to regulatory requirements and traditional high-priority concerns of the industry. The research agenda is developed through a process of consultation with participants and drinking water professionals. Under the umbrella of a Strategic Research Plan, the Research Advisory Council prioritizes the suggested projects based upon current and future needs, applicability, and past work; the recommendations are forwarded to the Board of Trustees for final selection. The foundation also sponsors research projects through the unsolicited proposal process; the Collaborative Research, Research Applications, and Tailored Collaboration programs; and various joint research efforts with organizations such as the U.S. Environmental Protection Agency, the U.S. Bureau of Reclamation, and the Association of California Water Agencies.

This publication is a result of one of these sponsored studies, and it is hoped that its findings will be applied in communities throughout the world. The following report serves not only as a means of communicating the results of the water industry's centralized research program but also as a tool to enlist the further support of the nonmember utilities and individuals.

Projects are managed closely from their inception to the final report by the foundation's staff and large cadre of volunteers who willingly contribute their time and expertise. The foundation serves a planning and management function and awards contracts to other institutions such as water utilities, universities, and engineering firms. The funding for this research effort comes primarily from the Subscription Program, through which water utilities subscribe to the research program and make an annual payment proportionate to the volume of water they deliver and consultants and manufacturers subscribe based on their annual billings. The program offers a cost-effective and fair method for funding research in the public interest.

A broad spectrum of water supply issues is addressed by the foundation's research agenda: resources, treatment and operations, distribution and storage, water quality and analysis, toxicology, economics, and management. The ultimate purpose of the coordinated effort is to assist water suppliers to provide the highest possible quality of water economically and reliably. The true benefits are realized when the results are implemented at the utility level. The foundation's trustees are pleased to offer this publication as a contribution toward that end.

Edmund G. Archuleta, P.E.
Chair, Board of Trustees
Awwa Research Foundation

James F. Manwaring, P.E.
Executive Director
Awwa Research Foundation

# ACKNOWLEDGMENTS

Projects of this nature depend heavily on utility participation to be successful. Only with the significant involvement of utility practitioners can we be certain that this research is relevant and useful to AwwaRF participant needs.

The authors of this report gratefully acknowledge the participation and funding from the following organizations: Cleveland Water, Columbus Water Works, Des Moines Water Works, East Bay Municipal Utility District, EPCOR, Honolulu Board of Water Supply, Metropolitan Water District of Southern California, Phoenix Water Services, San Diego Water, San Francisco Public Utilities Commission, Seattle Public Utilities, Tucson Water, and Winnipeg Water. In addition to these utilities that participated in the initial benchmarking consortium, three additional benchmarking partners were included in the case studies. We also acknowledge and thank these organizations, including the Regional Municipality of Halton, an unnamed electric, gas, and water utility, and an anonymous leading high tech worldwide manufacturing company. As attested to in the participant's code of conduct (Reference: APQC, Houston, TX), we respect the privacy of these organizations in sharing their information anonymously.

The advice and help of the Awwa Research Foundation project manager (Jennifer Warner) and the Project Advisory Committee (A.J. Henry, Milwaukee Water Works; Ed Blundon, Phoenix Water Services; Eric Hoopes, Intuitech; and Kiran Kothari, Gas Technology Institute) is especially noted with thanks and appreciation. Their commitment to this project, and to the industry, benefits us all.

EMA researchers included Terry Brueck, principal investigator, with Melanie Rettie, Bruce McClung, Rich Lackmann, Alicia Crumpton, and additional research assistance from Dr. Linda Paralez (DeMarche Consulting) and Rosy Spraker (Gartner Group).

# EXECUTIVE SUMMARY

Increasing costs. Reduced revenue. Increasing regulations. Changing workforce. Increased customer expectations. Constantly changing technology. These are only some of the challenges faced daily by utility managers. Technology offers one of the significant levers that utility managers have available to truly "do more with less." But it often is difficult for managers to know how much to spend on technology, where the expenditures should be, and what the expected return on that investment should be. What practices should be employed to make these investments truly effective? This research project addresses those questions.

This study provides utility and information technology (IT) managers guidance on: "How effectively am I delivering and managing my information technology?" A methodology is provided herein for answering that question, and establishing an ongoing improvement program. The initial objective of this project was to determine successful approaches for implementing utility IT systems given the range of factors that influence the utility IT environment today. This study cites many leading practices that create effective IT solutions.

There is no one "best" piece of technology that utilities should use. Hardware, software, and communications technologies are continuing to change at a rapid pace, and will continue to do so for the foreseeable future. This project is not intended to address technology-specific questions such as the "best" Laboratory Information Management System (LIMS) or the "best" relational database, or "best" networking technology. Technology solutions should reflect the specific utility context (size, other systems in place, employee skills, available funding and so on). What does endure are the management and business practices around how technology is selected, implemented, used, and managed. Those practices are process-oriented, repeatable, documented, and measurable: they are the focus of this project.

Nineteen organizations participated in this benchmarking project, including the initial thirteen water or combined water utilities of the benchmarking consortium plus three additional benchmarking partners and three research organizations. A benchmarking consortium of utilities was formed to initiate this research project, drawing on utilities that included leading practitioners in many IT areas, all of whom wanted to improve their IT effectiveness. The utility organizations represented a broad cross-section and variety of sizes, governance structures, services provided, and technology complexity. Their participation was key, and they provided active project direction and feedback during the course of the study. Two face-to-face workshops were conducted, supplemented by extensive surveys, web-conferences, team website interactions, and other remote communications methods. Case studies included two utilities from the initial consortium plus three benchmarking partner organizations.

In addition to the direct utility participation and guidance, the project team conducted extensive secondary research. The secondary research spanned all industries, not just the water utility business. Business practices regarding the use of technology are independent of the industry – much can be learned from other sectors such as manufacturing, telecommunications and banking. The specific technologies used will change, but the business practices that guide the selection, deployment, implementation and overall management of technology are more enduring. Utilities that adopt and implement effective IT practices will gain far more benefit from their IT investment regardless of the specific technologies deployed. This research was distilled into an initial set of effective IT business practices. These effective IT business practices were organized into eight focus areas, as depicted in the IT Framework shown in Figure ES.1.

Each of these focus areas is comprised of multiple elements, with each element containing multiple leading practices.

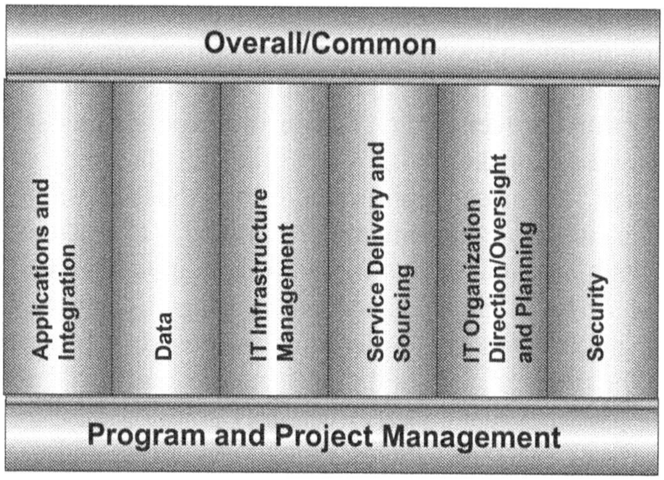


**Figure ES.1. Comprehensive IT framework addresses all IT business practice areas**

The following four steps summarize the overall project approach:

- **Plan**: Secondary research identified IT practice elements and grouped them into focus areas that significantly contribute to the effectiveness and value of IT services for a water utility. Utility participants helped frame and refine the focus area elements to prepare for subsequent survey tools. Potential benchmarking partners were identified and code of conduct agreed to by all participants.
- **Collect data**: For each focus area, specific practice elements and metrics were developed to identify leading practices. A survey instrument included business practice questions and metrics for the selected IT focus areas. The survey was distributed to the participating utilities, and data collected electronically.
- **Analyze data**: The completed surveys were analyzed for responses to the practice-related questions to identify case study candidates. Metrics and metric ratios were analyzed to validate practice question responses. For some focus areas, additional benchmarking partners were included to identify leading practices. The case study candidates were then further screened and qualified via phone interviews. Once qualified, site visits were conducted and the findings documented.
- **Report effective practices**: The detailed survey information, analyses, conclusions, and case studies have been documented and shared in detail with utility participants. Participants have used these results to create action plans for their own organization improvements.

A two-part survey was created to include questions on leading practices with metrics that relate to or support those practices. The first part of the survey addressed the metric information such as size of utility, number of customers served, services provided, dollars spent on technology, staffing, and so on. The second portion of the survey addressed how technology is prioritized, selected, deployed, monitored, and managed; the "practices" portion of the survey. The participating utilities provided guidance regarding those areas in which they would like advice, leading to development of practice questions for those elements. In all, 197 practices were surveyed. Allowable survey answers were structured into one of five possible categories, revealing information about the IT practice. The possible categories were frequency, formality, emphasis, results, level of automation, and extent of the use of the practice. The metric data was analyzed for potential correlation to the practices, and to discover new leading practices. For example, the IT budget as a percent of total revenue was analyzed, to discover what practices might enable lower expenditures while providing better results. Each participating utility completed the survey with additional benchmarking partners solicited after the initial survey results.

The survey results were analyzed for each utility by each focus area. The analysis indicated that some of the participating utilities had high scores in certain areas; these utilities were considered as potential case studies in those focus areas. Additionally, more than twenty external organizations were solicited as benchmarking partners to be additional case studies. Potential case studies were validated via the two-part survey and phone interviews, and for selected cases followed up by on-site visits and first-hand observations. A total of five case studies resulted from this winnowing process – two of them from the participating utilities. A total of 22 leading practices exhibited by the case studies are described in this report. Some practice areas contributed by the case studies include: alignment with strategic and tactical plans; application development and enhancements; budgeting; customer satisfaction; data management; governance and policy setting; performance management and project management. Additionally, other participating utilities contributed potential leading practices.

The survey results of the case studies were compared to the other participating utilities. A number of observations and possible implications were drawn from the findings. In general, the average scores of the case studies were notably higher than the average scores of the participating utilities (4.2 versus 3.1, on a 5 point scale), indicating that the case study organizations do derive greater results from their technology investments. Some other key differences were noted between the case studies and other respondents. In general, the case study organizations:

- Have a higher total annual revenue per customer for all services provided
- Have a lower IT budget as a percentage of total operating budget
- Have a higher percentage of internal IT staff (versus external contractors)
- Have a higher training budget as a percentage of total IT budget
- Support fewer PCs per IT staff

Several possible implications and additional research hypotheses could be drawn from these observations including the extent to which investments should be made in training, staff levels, and redesigning internal practices, compared to hardware and software investments. Some of these key implications include:

- Leading practices in IT can be achieved in an organization with sufficient but not excessive IT spending. The leading practices organizations spent less than 5 percent of their operating budget on all IT related costs.
- When compared to other organizations, leading practice organizations may spend more for sufficient and skilled IT staff and reduce spending on other IT resources (such as hardware and software). The leading practice organization's IT staff averaged almost 10 percent of the total company full-time equivalents (FTEs).
- Leading practice organization's demonstrated a clear focus on changing or redesigning their internal business practices using existing tools to gain efficiencies. This required that the IT staff provide more than just technical prowess, but also provide competencies and capabilities such as business understanding, effective communications, negotiation skills, meeting facilitation, and sound project management.

A number of tools for assessing and improving utility information technology were developed as part of this project. Each of these tools can be used directly by utilities as a means of self-assessment, evaluation of practices for improvement, and action planning.

- Tools to Assess: IT practices and metrics survey. This survey provides a snapshot that when applied over time (e.g., an annual survey), indicates the trend of practices improvement at the utility. The survey instrument is not included in this written report.

- Tools to Evaluate: IT practices maturity grid. This grid extends the concepts of the Software Engineering Institute's (SEI) Capability Maturity Model®. It provides a basis for selecting particular technology practices upon which to focus, and provides initial guidance regarding how to improve those practices. Examples of leading practices from the case studies can be used to evaluate possible improvements.

- Tools to Plan: action planning and business case worksheets. These worksheets further the assessment results and set the stage for action. The action planning worksheet turns the output from the IT practices maturity grid into an actionable practices improvement plan to be implemented at the utility. The business case worksheet provides guidance to managers as they identify the benefits to be obtained from improving their IT practices.

Ensuring that value is received from your investment in information technology requires constant attention, and skilled management. A number of challenges and obstacles exist. Some of these are identified and discussed below.

- Information technology is a very immature and young field – just barely 50 years old. This has many implications, including the ongoing and continuous development of new technologies. Faster hardware, better software, wider communications pipelines, smaller and more flexible form factors, and new capabilities are developed every year. This causes misguided thinking that it is the technology that is important, rather than how technology is employed, and the changes that technology enables. People often put their focus on keeping up with hardware and software changes, rather than focusing on the real issue, which is how to get benefits from whatever hardware and software is deployed.

- When new capabilities emerge, new practices must be developed – there is no guideline for leading practices when something hasn't existed before. Hence there is a discovery period before it is clear what practices are effective. Additionally, there is no good mechanism for finding, or sharing, effective practices. It is largely an environment in which individual organizations and utilities are left to discover and create their own practices. This takes a great deal of time and effort.
- It is very easy to underestimate the cultural changes required to adopt and institutionalize new ways of working. To truly adopt new practices requires much more than just a new procedure – it requires training, communication, feedback, persuasion, leadership, vision and commitment.

Because the science of information technology is still developing so rapidly, "leading" or "best" practices will also continue to rapidly develop. It is important to continue questing for the most effective means of ensuring that utilities receive value for the IT investments they make.

# CHAPTER 1
# INTRODUCTION AND APPROACH

## PROJECT GOAL

All utilities spend money on information technology, including the maintenance of existing systems, procurement and implementation of new systems, labor, and consulting fees. Many utilities spend millions of dollars per year on information technology. Yet most utility Managers and IT Managers question whether they are getting value for their investments. They struggle with decisions of how much to invest, where to invest it, and what payback should be expected.

The purpose of this study is to provide guidance to utility managers and IT managers as they address the following question: "How effectively am I delivering and managing my information technology?" The initial objective of the project was to determine successful approaches for implementing utility IT systems given the range of factors that impact the IT and utility environments. The stated question was the researchers' and participants' method to focus the research on successful practices for these utility IT systems.

## PROJECT APPROACH

"Information Technology" is an extremely broad term. It includes dimensions of hardware, communications, operating systems, software applications, data and how it is organized, business rules, people, training, and so on. To provide structure to the analysis, an IT Practices Framework was developed. This framework provides a comprehensive view of key information technology components, including aspects of organization and work practices. It consists of eight information technology focus areas as shown in Figure 1.1. The focus area framework includes "horizontal" elements that impact each of the "vertical" practice elements.

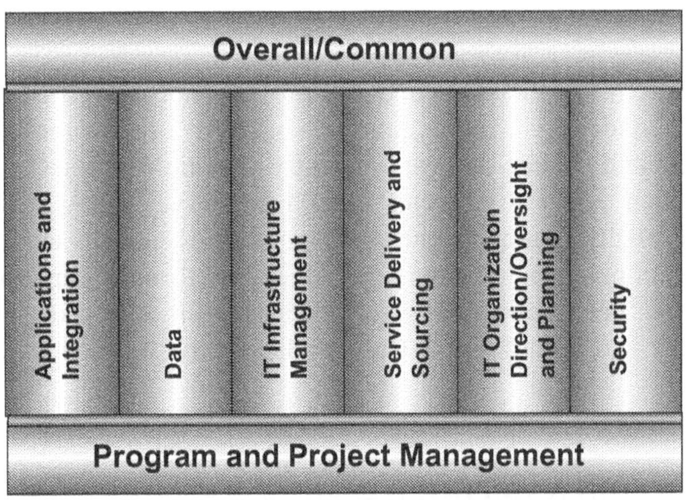


**Figure 1.1: Comprehensive IT framework addresses all elements**

Each focus area consists of several "practice elements" that address the business and management practices within the focus area. The eight focus areas are comprised of a total of 52 practice elements. Two focus areas (overall/common, and program and project management), impact all other focus areas, thus they run horizontally across the diagram. The focus areas are summarized in Table 1.1, and more fully described in Appendix F.

**Table 1.1**
**Focus areas and practice elements**

| | Focus Area | Practice Element |
|---|---|---|
| 1 | Applications and Integration | Software applications; application development and enhancement; application upgrading and decommissioning; e-Business; application-specific training and support; and integration (including tools and procedures) |
| 2 | Data | Data definition, models and standards; data acquisition/input; data business rules; data storage; data analysis and reporting; data integration; data transformation and conversion; data management; data stewardship; and data quality |
| 3 | IT Infrastructure Management | IT asset tracking; network management; telephony; networks; wireless; IT architecture; and IT service desk |
| 4 | Service Delivery and Sourcing | Procurement policies and procedures; service delivery alternatives; service level management; and systems development life-cycle and management |
| 5 | IT Organization and Direction/ Oversight and Planning | Governance and policy definition; strategic/organizational plans; IT vision, mission, goals; budgeting (capital, operations and annual); total cost of ownership; business continuity/disaster recovery; roles and responsibilities; and business process redesign |
| 6 | Security | Policies and procedures; and tools |
| 7 | Program and Project Management | Program definition and management; alignment with strategic and tactical plan; project planning, management and tracking; quality control/quality assurance; and testing strategies |
| 8 | Overall/Common | Hiring; performance management; IT staff development and training; organizational structure and reporting customer satisfaction; organizational communications; IT organizational performance and continuous improvement; technical research; and marketing |

Each focus area (e.g., applications and integration, data, IT infrastructure management) was further investigated to identify "leading practices" used by organizations to provide effective IT solutions, both within and outside the water industry. The utility project participants provided guidance regarding which practice elements should be surveyed. A survey was created to identify the current IT practices among participating utilities, and to compare those practices with the identified leading practices. The survey included performance metrics that validated the leading practices that had been defined. It also indicated utilities and organizations employing the leading practices that had been defined. Case studies of those practices are provided later in this report.

## UTILITY PARTICIPATION AND PROFILE

The utility participants were key throughout the project. This consortium-based project included a collaborative effort of 16 leading water utilities and other organizations from throughout North America. The participating utilities were directly involved in the design and direction of the study. The involvement of leading utility participants, complemented by extensive research of the literature and out-of-industry benchmarking, provides the drinking water industry with current leading practices and case study examples for implementing effective IT solutions.

As a beginning step in designing the study, profiles of the participating utilities were completed, providing an initial baseline from which to work. Each utility provided information about their ownership/governance, utility services provided, number of customers, annual revenue, and so on. The participants represent a broad cross-section of utility organizations as shown below.

- A variety of ownership and governance structures were represented. Most participants were part of a local governance such as city or county. Others were either part of an independent structure (commission, authority, district, etc.) or private.
- A variety of utility services provided – all utilities provide potable drinking water transmission. Most also provide potable drinking water treatment. About half of the utilities wholesale raw water, as well as providing wastewater collection, treatment and reclaimed water services. Additional services provided by the participants included electric services, solid waste collection or billing, hydroelectric generation, and natural gas retail services. One utility also actively provides and markets consulting services.
- The number of customers ranged from about 68,000 to 1,300,000.
- Annual revenue ranged from approximately $26,000,000 to $1,400,000,000.
- Annual IT operating budget ranged from approximately $268,000 to $21,000,000.
- Staffing – total utility employees ranged from 205 to about 3,500.
- Staffing – total IT employees ranged from 4 through about 110.

In addition to providing profile information, participants provided input on their most significant or recent IT accomplishments, the biggest IT issues/obstacles they face, and specific project expectations. Summaries of their perspectives are provided in Appendix F.

The participant input helped to further define the project focus. Providing each utility with a summary of responses also created a vehicle for dialogue amongst the participants to understand the enormity of what the water/wastewater industry has recently accomplished and what opportunities lay before them in the area of information technology.

## PROJECT FOCUS

The first of two project workshops was held to develop a shared understanding of major IT trends and issues, and to focus the project based on input from the participating utilities. The IT Framework described above was an output of this workshop (shown previously in Figure 1.1). This framework was exercised and validated during the remainder of the project.

## CONSORTIUM APPROACH

Consortium benchmarking is a recognized and effective method of discovering leading practices. It allows project participants to directly participate and guide the design, direction, and outcomes of the study. The methodology provides for participant interaction throughout each step. Figure 1.2 (Brueck et al. 2002) shows how the utilities were involved during key project steps and tasks.

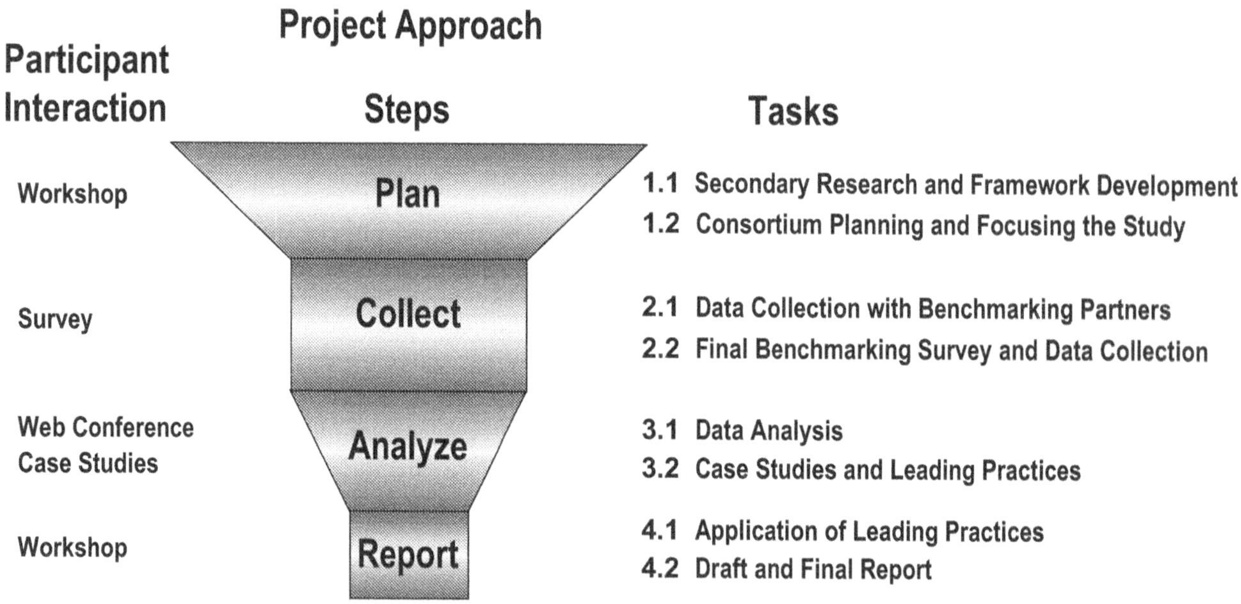


**Figure 1.2. Consortium benchmarking approach to projects** (Adapted from Brueck et al. 2002)

It should be noted that all participating utilities were asked to read, understand and commit to the Benchmarking Code of Conduct developed by APQC (Reference: APQC, Houston, TX). This is a document that defines the guidelines and ethics that benchmarking organizations agree to follow.

Project Team members and utility participants all signed a collective statement indicating their commitment to the Code of Conduct.

## SECONDARY RESEARCH

The project team and the participating utilities brought great experience to this project. To complement this experience, the project team conducted a survey of the literature "secondary research" (information collected from other published sources). This additional in-depth secondary research was conducted for each of the focus areas in order to discover applicable leading practices. This information was used as much of the basis for the survey questions. A bibliography of the secondary research is included in Appendix H.

The secondary research efforts were greatly enhanced by the partnership of Gartner Group, Inc. Gartner provided global expertise and experience across multiple industry and business sectors. Their library of information technology research was made available to the Project Team. A summary of Gartner's presentation on "IT Trends and Drivers" is included in Appendix D.

# CHAPTER 2
# SURVEY PROCESS AND FINDINGS

An electronic survey instrument was developed to identify and assess current IT practices, as organized and focused by the participating utilities. This chapter describes the survey process itself, the analysis of the survey findings, the use of survey data to select case studies and benchmarking partners, and the results of a supplemental security survey.

## FINAL BENCHMARKING SURVEY

Participating utilities provided feedback on the development of the survey. The survey emphasized practices that participants identified as issues in the implementation of successful IT solutions. After incorporating the feedback, two utilities piloted the survey process and provided additional feedback. The full survey is included in Appendix A.

The final survey was comprised of two parts: utility profile (or metric) data, and utility information technology practices questions. The metric data was analyzed for potential correlation to the practices. It was also analyzed to provide a comparison for discovering new leading practices. For example, the IT budget as a percent of total revenue was analyzed, to discover what practices might enable lower expenditure with higher results. In addition, some of the questions included metrics to try to verify or validate the practices.

Thirty-nine of the 52 practices elements were addressed in the survey, based on the participating utilities' priorities. The practice elements not included in the survey were generally deemed to be less of a priority by the participating utilities. All eight focus areas were included. Ultimately, a total of 197 survey questions were crafted regarding current practices. Each survey question was framed to provide information in one of the following categories:

Frequency: How often is the business practice used?
Formality: How well established is the business practice within the organization?
Emphasis: What aspects of the practice are most important to the organization?
Results: How much benefit has the organization received from the practice?
Systems: What level of automation, or development exists in the system being employed?
Percentage: How extensively is the business practice is currently being applied?

The possible responses for each question were standardized and ranged from a score of 1:
Never do this
No formal approach
Not emphasized
No results seen
No system exists

to a score of 6:
Always do this – standard operating procedure
Heavily emphasized – one of our business principles)
Unparalleled results – a total success
State of the art system is in place and meets all users needs

Below are samples of 'Emphasis'and 'Formality' practice questions:

**PRACTICE: A flexible reward system is in place for IT employees.**

| A.1. A flexible reward system is in place that may include recognition, training, and other non-monetary rewards. | |
|---|---|
| Not emphasized –1 | ○ |
| Modest emphasis; some efforts underway – 2 | ○ |
| Moderately emphasized; something we try to adhere to – 3 | ○ |
| Generally emphasized; something we do and check – 4 | ○ |
| Strongly emphasized; something we measure/reward – 5 | ○ |
| Heavily emphasized; one of our business principles - 6 | ○ |

**PRACTICE: Performance Management is an ongoing process throughout the year for all IT employees.**

| A.2. There are multiple formal feedback sessions throughout the year (more than one time per year). | |
|---|---|
| No formal approach – 1 | ○ |
| Formality as needed; based on criticality – 2 | ○ |
| Sometimes formal procedure: an aspect of some situations – 3 | ○ |
| Typically formal; an aspect of most situations – 4 | ○ |
| Often do this, omitted only in exceptional circumstances – 5 | ○ |
| Always do this: standard operating procedure – 6 | ○ |

## CONDUCTING THE SURVEY

The final survey was distributed for completion by the thirteen utility participants. Quality control for the practices responses was addressed by four factors:

1. Selecting a cross-functional team of perspectives (including IT users [utility managers and staff perspectives] and IT staff [managers and IT practitioners]) to address each question and come to consensus.
2. Including a utility subject matter expert who understood the question being asked.
3. Using a robust rating system (previously described) for each practice.
4. Use of consistent terms to help respondents interpret terminology in the same way.

It was recommended that each participating utility use the process shown in Figure 2.1 to complete the practices portion of the survey.

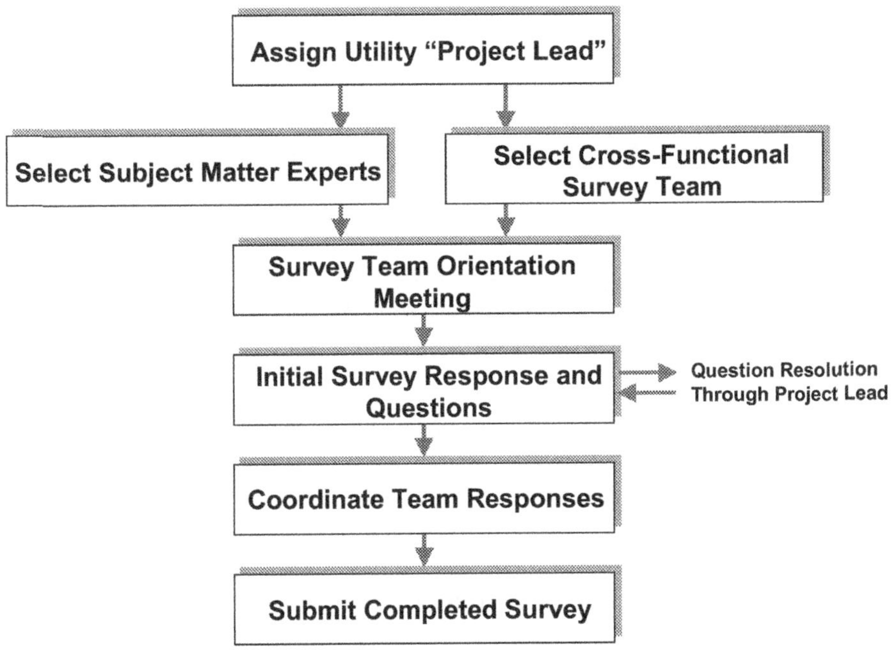


**Figure 2.1. Process flow for completing information technology survey**

The Project Team provided support for the utilities as they completed their surveys to ensure that the questions were understood. The completed surveys were reviewed and analyzed. In some cases, follow-up was needed to ensure consistency of profile data. Key findings from the analysis were presented to the utility participants at the second and final workshop. The participating utilities comments and suggestions were incorporated into this report.

## SURVEY PART I: METRICS/PROFILE

The metric/profile data provides quantifiable measures or statistics depicting the overall business enterprise and its information technology environment. For example, it contains information about the organization's ownership, business services provided, customers, revenue, budget, staffing, training, information technology capital expenditure history, and levels of satisfaction regarding information technology.

Utilities were asked to provide one of three qualitative measures regarding the accuracy level of their metric/profile data. These measures of accuracy were taken into consideration by the project team when analyzing the data:

***A*** (Actual Data): collected the way the question is asked
***E*** (Estimated Data): calculated in order to respond to the question
***R*** (Rough Guess): a Subject Matter Expert's best guess

Below is a sample of the metrics/profile portion of the survey. Results from the metric/profile portion of the survey can be found in Appendix B.

**C. UTILITY CUSTOMERS:**

| Potable Water Customers: | (Number of Accounts) | Year X | Year Y | Year Z | 3 Year Avg. | A=Accurate; E=Estimated; R=Rough Guess |
|---|---|---|---|---|---|---|
| | | | | | | A / E / R |
| 1 | Wholesale #: | ______ | ______ | ______ | ______ | O O O |
| 2 | Commercial/Industrial #: | ______ | ______ | ______ | ______ | O O O |
| 3 | Residential #: | ______ | ______ | ______ | ______ | O O O |

The overview charts that follow (Figures 2.2–2.7) provide a summary profile of participating utilities. The utilities represent a wide cross-section of utility sizes, services provided, IT funding levels, and customers served.

Note that there are blanks in some of the charts. This is due to respondents either not providing data or the area was not applicable to their organization.

Numbers 1 through 11 represent participating utilities. Letters A through E represent the case studies. Note that two of the case studies were also participating utilities.

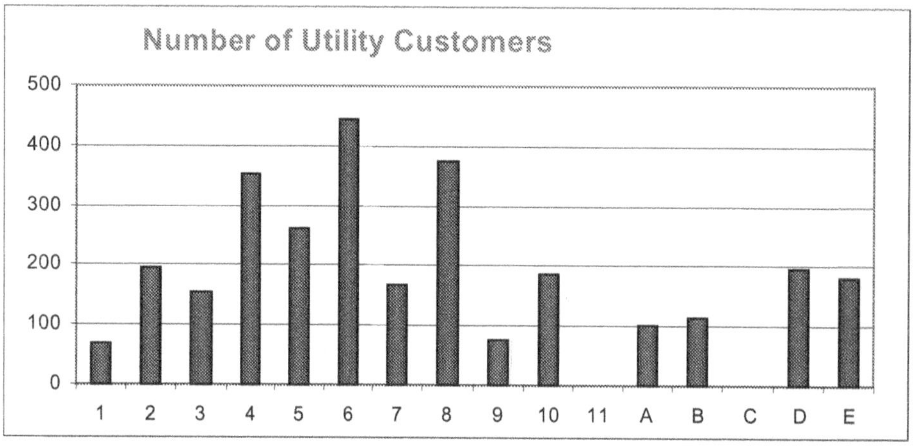


**Figure 2.2. Number of utility customers**

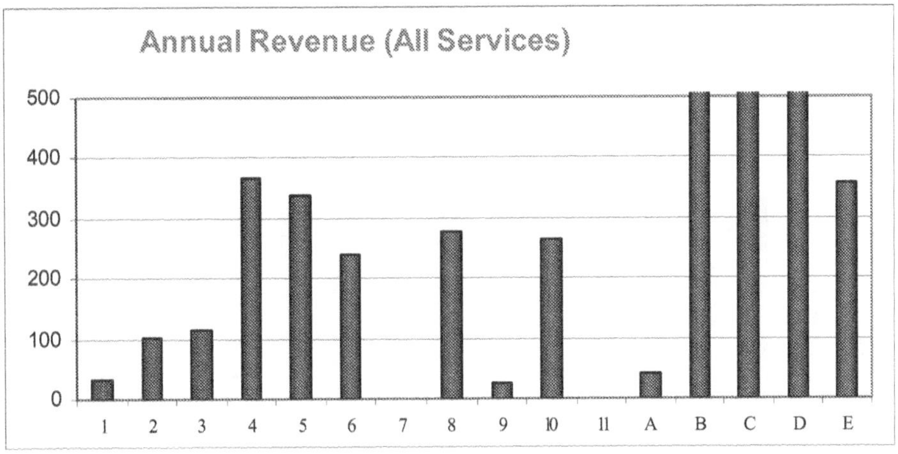


**Figure 2.3. Annual revenue, all services**

**Figure 2.4. Total number of employees**

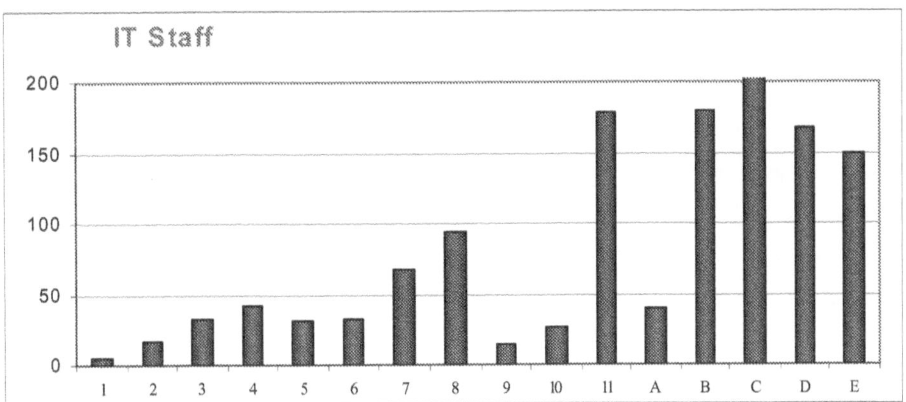


**Figure 2.5. Total number of IT staff**

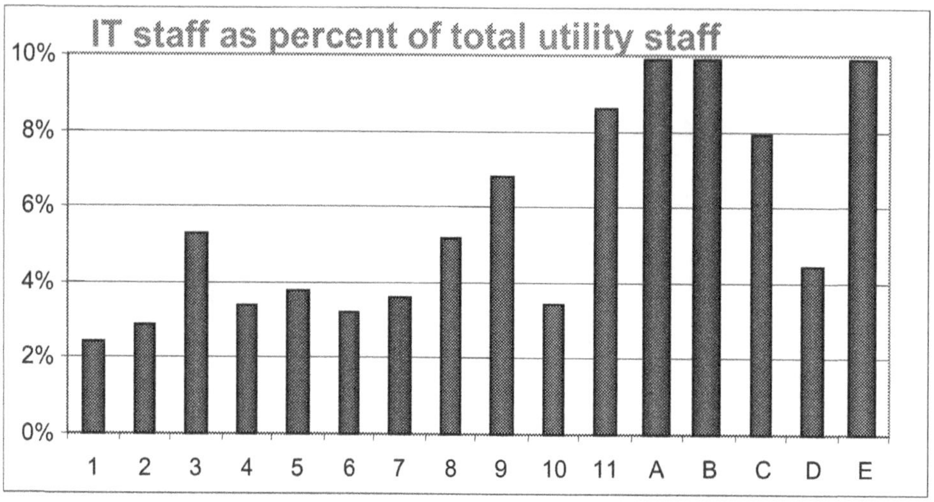


**Figure 2.6. IT staff as percent of total utility staff**

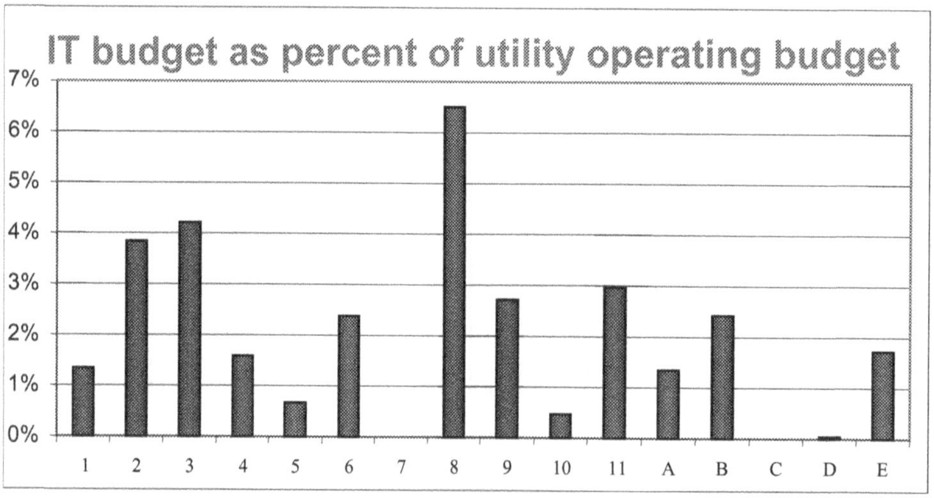


**Figure 2.7. IT budget as percent of utility operating budget**

## SURVEY PART II: PRACTICE QUESTIONS BY FOCUS AREA

### Focus Area Averages

Figure 2.8 is an overview of the utility responses to all of the focus areas. The individual responses to questions in each focus area were averaged, to provide a summary view of a particular organization's practices in that area.

The areas highlighted in light gray indicate a response of 4 or above and the darker gray highlighted areas represent top leading practices with responses of 5 or better. The average case study score was noticeably higher than the non-case study scores (4.2 vs. 3.1, on a five point scale). Individual charts showing responses for all practices questions are provided in Appendix B.

*See section on Final Benchmarking Survey (page 5) for explanation of scoring.*

Legend: 4.0 to 4.9 (light gray); 5.0 and greater (dark)

| | | | Participating Utilities | | | | | | | | | | Case Studies | | | | | Group Averages | | |
|---|---|---|---|---|---|---|---|---|---|---|---|---|---|---|---|---|---|---|---|---|
| | | | 1 | 2 | 3 | 4 | 5 | 6 | 7 | 8 | 9 | 10 | A | B | C | D | E | ALL | Total 1-10 | Total A-E |
| Focus Area: | A | Overall/Common | 1.5 | 2.6 | 3.2 | 2.3 | 2.0 | 2.7 | 2.9 | 4.4 | 3.5 | 3.3 | 4.2 | 4.5 | 5.4 | 4.2 | 3.3 | 3.3 | 2.8 | 4.3 |
| Focus Area: | B | Applications and Integration | 2.5 | 3.0 | 3.5 | 2.6 | 3.0 | 3.3 | 3.3 | 4.7 | 3.3 | 3.0 | 4.3 | 4.0 | 5.0 | 4.5 | 2.8 | 3.5 | 3.2 | 4.1 |
| Focus Area: | C | Data (Including Records Management) | 2.1 | 2.5 | 2.7 | 2.8 | 2.8 | 2.9 | 2.4 | 4.8 | 2.6 | 2.0 | 3.5 | 2.6 | 5.0 | 3.9 | 2.1 | 3.0 | 2.8 | 3.4 |
| Focus Area: | D | Infrastructure Management | 2.2 | 3.0 | 3.5 | 3.0 | 2.5 | 3.6 | 2.4 | 4.7 | 3.8 | 3.1 | 4.7 | 2.9 | 5.6 | 4.7 | 4.2 | 3.6 | 3.2 | 4.4 |
| Focus Area: | E | Service Delivery and Sourcing | 2.0 | 2.0 | 3.5 | 2.0 | 3.3 | 3.7 | 2.7 | 4.2 | 2.6 | 1.8 | 3.5 | 2.1 | 5.5 | 5.5 | 4.1 | 3.2 | 2.8 | 4.1 |
| Focus Area: | F | IT Organization, Direction, Oversite and Planning | 2.4 | 2.9 | 3.5 | 2.8 | 3.8 | 4.0 | 2.7 | 5.0 | 3.9 | 3.3 | 4.4 | 2.6 | 5.4 | 4.5 | 4.8 | 3.7 | 3.4 | 4.3 |
| Focus Area: | G | Program and Project Management | 2.5 | 3.1 | 4.3 | 2.3 | 3.7 | 4.5 | 2.7 | 4.5 | 3.2 | 2.6 | 5.0 | 2.1 | 5.2 | 5.0 | 5.0 | 3.7 | 3.3 | 4.5 |
| Focus Area: | H | IT Security | 3.3 | 3.4 | 3.9 | | 3.6 | 4.3 | | 4.5 | 4.4 | 3.2 | 3.5 | | 5.5 | 4.1 | | 4.0 | 3.8 | 4.4 |
| | | AVERAGE OF ALL FOCUS AREAS: | 2.3 | 2.8 | 3.5 | 2.5 | 3.1 | 3.6 | 2.8 | 4.6 | 3.4 | 2.8 | 4.1 | 3.0 | 5.3 | 4.5 | 3.8 | 3.7 | 3.1 | 4.1 |
| | | AVERAGE OF FOCUS AREAS A-G (Excludes IT Security): | 2.2 | 2.7 | 3.4 | 2.5 | 3.0 | 3.5 | 2.8 | 4.6 | 3.3 | 2.7 | 4.2 | 3.0 | 5.3 | 4.6 | 3.8 | 3.7 | 3.1 | 4.2 |

**Figure 2.8. Utility responses show a range of practices across focus areas**

Focus Area A, Overall/Common, addresses overall and common IT issues. This includes organization structures and staffing approaches, as well as performance management/ measurement. In addition, most organizations have common methods and tools (e.g., organizational communications and customer feedback mechanisms) that may be applied to any of the subsequent focus areas. Elements in this focus area include:

- Hiring
- Performance Management
- IT Staff Development and Training
- Organizational Structure and Reporting
- Customer Satisfaction
- Organizational Communications
- IT Organizational Performance and Continuous Improvement
- Technical Research
- Marketing

Figure 2.9 below shows the organization scores for this focus area.

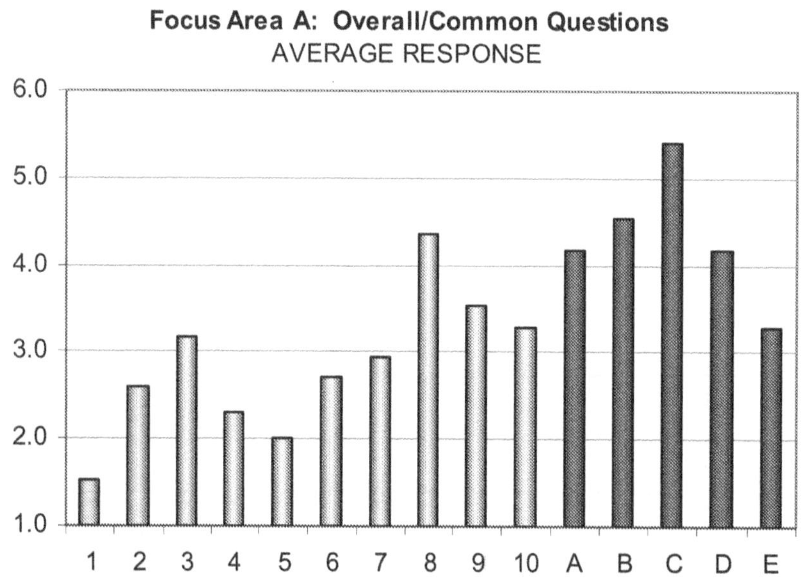


**Figure 2.9. Focus Area A: Overall/Common questions**

Focus Area B (Figure 2.10), Applications and Integration, addresses issues related to typical utility IT applications, including business and administrative systems, as well as systems to support operations and maintenance functions. Successful implementation of the same application (even the same vendor) varies from utility to utility. Continued emphasis on integration and interoperability of applications will be key to effective IT solutions. Most organizations have established basic application development methodologies and corresponding specific development tools. This focus area also deals with the integration of many, if not all, IT business applications and services with other applications, as well as with Internet (World Wide Web), Intranet, and browser technology. Elements in this focus area include:

- Software Applications
- Application Development and Enhancement
- Application Upgrading and Decommissioning
- e-Business
- Application Specific Training and Support
- Integration (including tools and procedures)

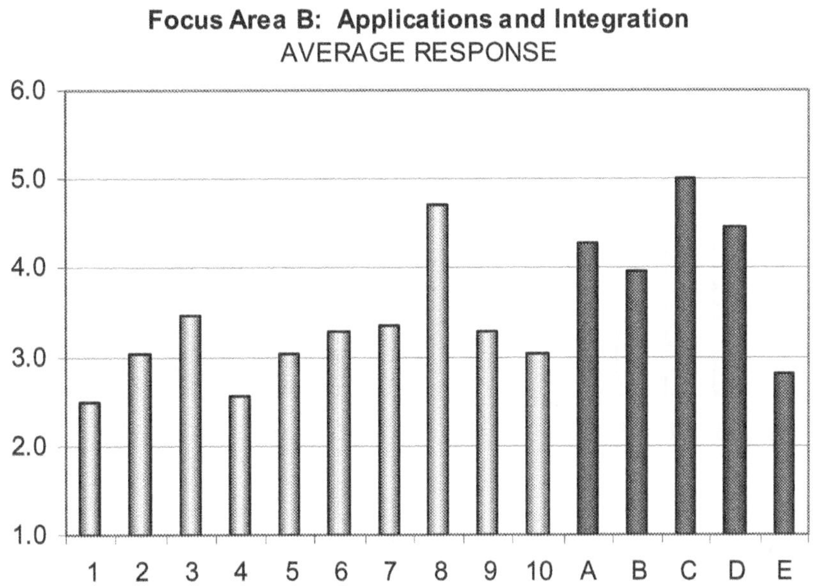


**Figure 2.10. Focus Area B: Applications and Integration**

Focus area C (Figure 2.11) deals with Data, including records management. Data is the raw material produced by one or more business processes that create and maintain it. It is both an output from business processes, as well as an input into business processes. Increasing data complexity and ambiguity, development of security policies, and increased need for integration between applications require increased rigor around data. Industry-wide agreement and standardization will continue to grow in importance due to increased use of supply-chain interactions and the cost advantage of uniform performance measurement/decision support systems. Focus elements include:

- Data Definition, Models, and Standards
- Data Acquisition/Input
- Data Business Rules
- Data Storage
- Data Analysis and Reporting

- Data Integration
- Data Transformation and Conversion
- Data Management
- Data Stewardship
- Data Quality

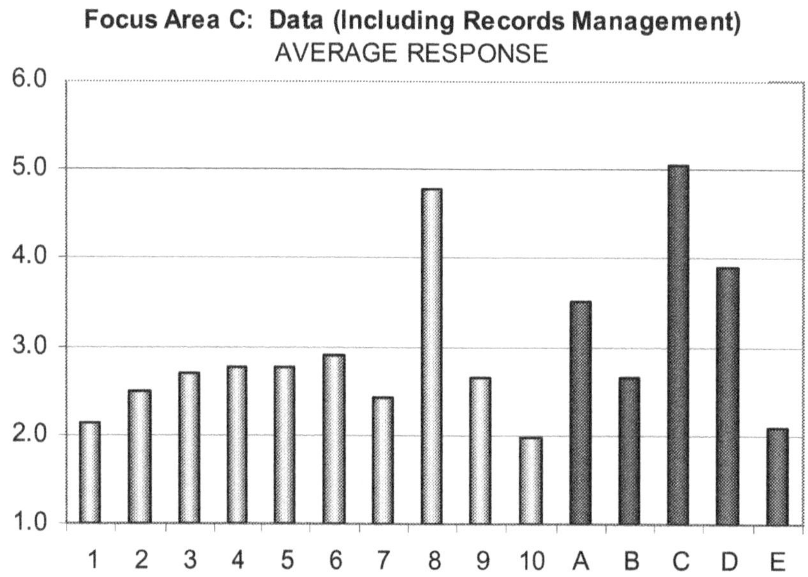


**Figure 2.11. Focus Area C: Data (including records management)**

Focus Area D (Figure 2.12) addresses Infrastructure Management. This broad focus area includes several elements and technologies: telecom, networking, wireless technologies, IT architecture, and security. It includes managing the IT infrastructure assets, as well as automated systems for real-time monitoring of network, server, and application systems. It also includes the integration of traditional and more newly developed business applications with telephony (voice telephone communications) systems. Common examples include the integration of database information with customer service applications, and the concept of "unified messaging" where voice-mail, e-mail, fax and audio-mail are integrated with each other. The networking element incorporates network media (cabling), hardware (switches, routers, firewalls, etc.), network management systems, and other components and subsystems required to support the IT environment. The wireless component includes the use of wireless (radio-based, infrared-based, laser-based) technology to augment wired network topologies. It is most commonly applied toward providing mobile access to Internet-based, and enterprise-based IT applications and services. The IT architecture encompasses different approaches such as thin client-server, N-tier (multi-leveled architecture), and web services. Security is critical to all of these elements, especially as the Internet becomes more heavily relied upon for enterprise business applications and networking. This IT element deals with protection of data, systems, and trusted relationships.

Focus elements include:

- IT Asset Tracking
- Network Management
- Telephony (IVR, ACD, Unified Messaging)
- Networks (Cabling, Routers, Firewalls)
- Wireless (Platforms and Integration)
- IT Architecture
- IT Service Desk (Help Desk, Desktop Management and Upgrades)

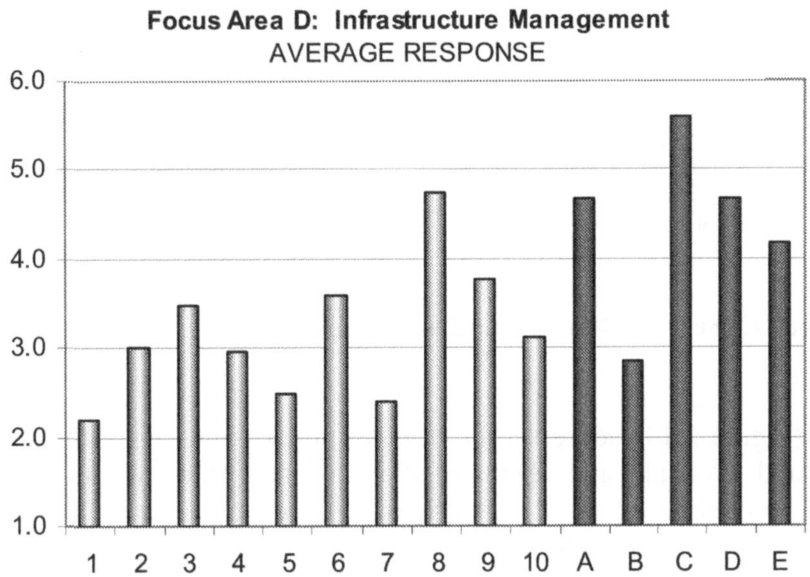


**Figure 2.12. Focus Area D: Infrastructure Management**

Focus Area E (Figure 2.13) addresses Service Delivery and Sourcing. Cost/benefit and risk assessment matrices have been developed to determine the tradeoffs in selecting service delivery and sourcing options. Service level and management level agreements are a critical aspect of this area in establishing requirements and the parameters around measuring the delivery of those requirements. This focus area also includes ongoing life cycle management of systems, whatever the delivery mechanism. Maintenance (both preventive and reactive) requires a systematic inventory of the environment (hardware, software, configurations, protocols) and continuous monitoring with an emphasis on identifying performance issues through "what if" scenarios associated with problem resolution, planned expansions, and application roll-out. Focus elements include:

- Procurement Policies and Procedures
- Service Delivery Alternatives
- Service Level Management
- Systems Development Life-Cycle and Management

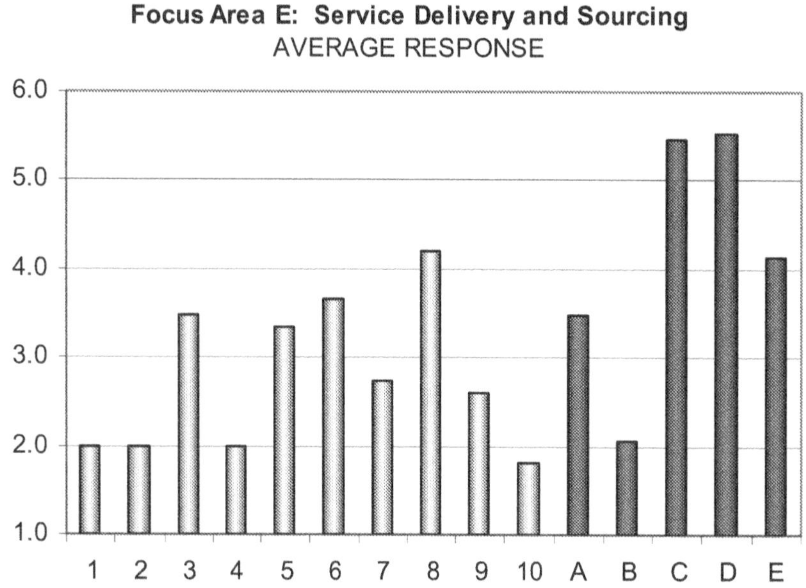


**Figure 2.13. Focus Area E: Service Delivery and Sourcing**

Focus Area F (Figure 2.14) addresses IT organizational issues, as well as the overall technology direction, and oversight and planning of IT. Clarity of organizational goals and policies is essential for providing effective IT services on an ongoing basis. This clarity must also address the roles and goals of multiple jurisdictions. Work practices must support overriding policy principles. Technology policies are derived from business objectives and are used to lower the cost of technology ownership through the decision-making processes. Strategic planning services address all aspects of information technology, including current situation assessments, communications technology vision development, strategic requirements development and recommendations. This includes the hard technical areas of business systems as well as business planning and staffing, organizational issues, and budget development to address change management. Planning efforts move from strategic to tactical to operational to budgetary. Focus elements include:

- Governance and Policy Definition
- Strategic/Organizational Plans
- IT Vision, Mission, Goals
- Budgeting (Capital, Operations and Annual)
- Total Cost of Ownership
- Business Continuity/Disaster Recovery
- Roles and Responsibilities
- Business Process Redesign

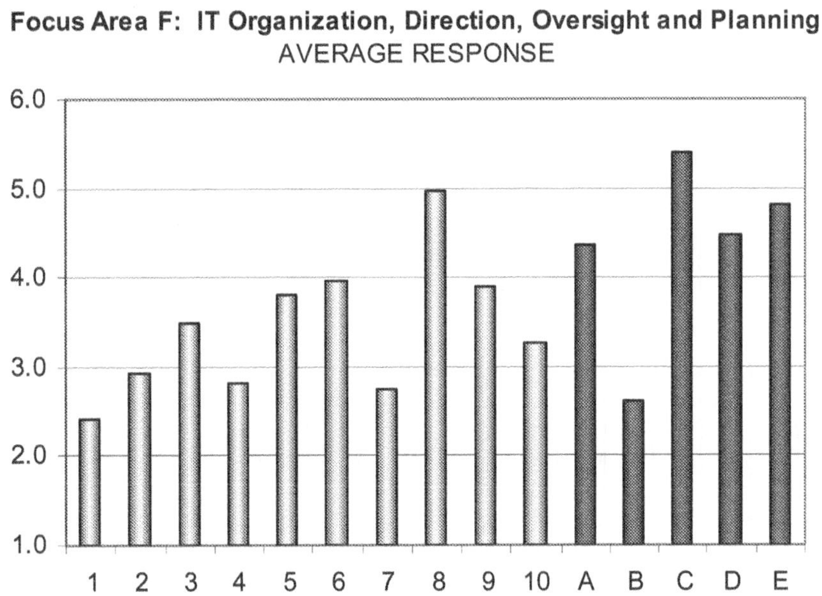


**Figure 2.14. Focus Area F: IT Organization, Direction, Oversight and Planning**

Focus Area G (Figure 2.15) addresses Program and Project Management. Programs are large projects with multiple subprojects. Projects are well-defined team efforts to achieve a goal that has a distinct beginning and end, milestones and budget. Projects require well-defined roles and responsibilities and methodologies establishing the activities throughout the entire life cycle. A key component of effective programs and projects is a quality program, including testing and validation. Elements in this focus area are:

- Program Definition and Management
- Alignment with Strategic and Tactical Plan
- Project Planning, Management and Tracking
- Quality Control/Quality Assurance
- Testing Strategies

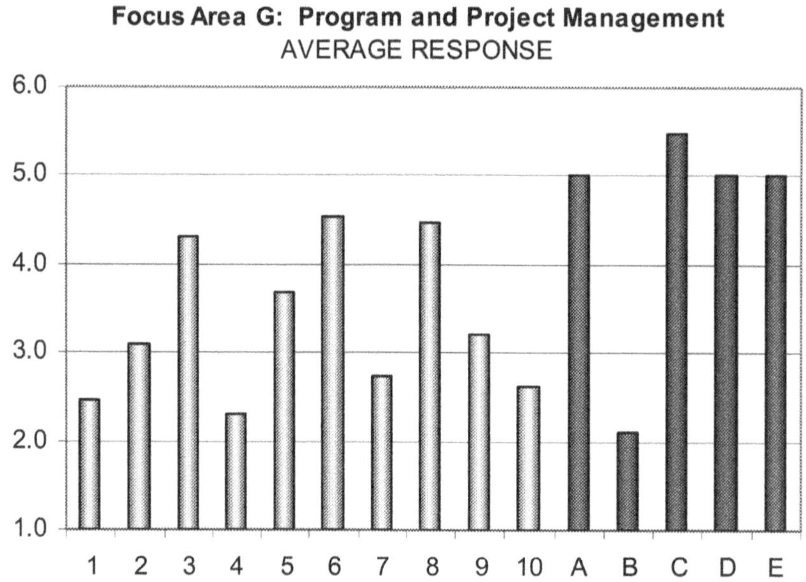


**Figure 2.15. Focus Area G: Program and Project Management**

Focus Area H (Figure 2.16) addresses security. Information technology security involves the protection of computer technology assets, including hardware, software, and data. IT security addresses the concepts of "CIA": systems and data Confidentiality, Integrity, and Availability. This involves making sure that only authorized users have access to systems and data; detecting and preventing unauthorized access and/or tampering with data (including virus detection, prevention and recovery); data encryption; hardening systems; and disaster recovery.

Hardening systems involves actions such as changing hardware and software factory default settings, and generally customizing default levels of security. From a cyber-security perspective, disaster recovery includes processes involved with backing up data, performing data restores, and rebuilding systems from backups.

- Policies and Procedures
- Tools

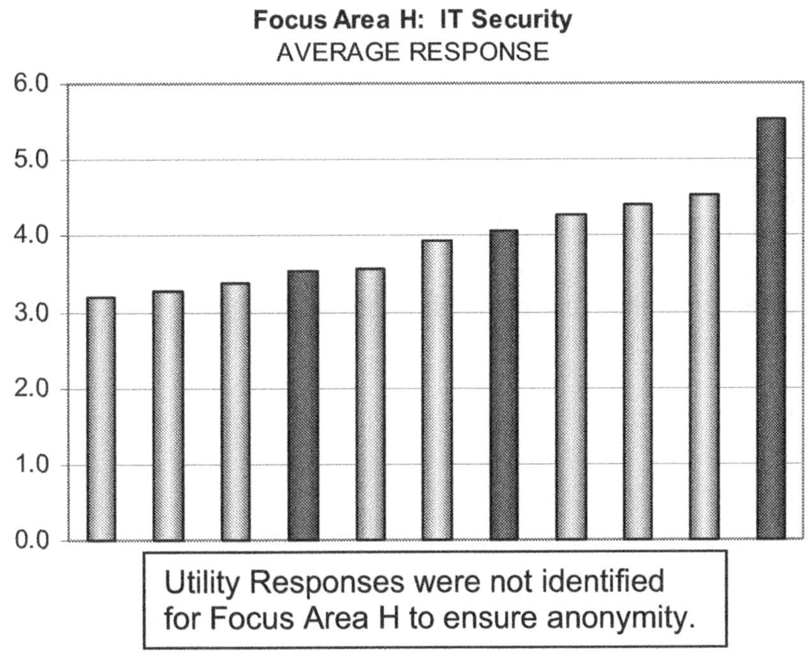


**Figure 2.16. Focus Area H: IT Security**

## RESPONSE ANALYSIS

A second interactive web conference was conducted to review the initial findings of the metrics and practices questions. The purpose of the web conference was to obtain feedback from the participants on the survey instrument itself, to confirm the process used to determine the responses, and to validate the responses themselves. A number of key ratios were graphed and selected for review, based on preliminary findings. These can be found in Appendix B – Survey Response Summaries.

The survey analysis uncovered some participating utilities with high scores in certain focus areas. These utilities were considered as potential case studies as shown in Figure 2.17.

| Focus Area | Potential Utility Case Study |
|---|---|
| Overall/Common | D, E |
| Applications and Integration | D, E |
| Data | D, E |
| IT Infrastructure Management | D, E, 8, 9 |
| Service Delivery and Sourcing | D, E |
| IT Organization Direction/Oversight and Planning | D, E, 3, 5, 8 |
| Program and Project Management | D, E, 3, 6, 8 |

**Figure 2.17. Participating utility responses indicated leading practices in each focus area**

## SELECTING CASE STUDIES AND BENCHMARKING PARTNERS

Responses from utilities D, E, and 8 indicated potential leading practices in three or more of the Focus Areas. The Project Team performed further analysis to validate the responses; ultimately utility D and E were selected to be included as case studies. Utility 8's data were unable to be validated, thus they were not considered for case study.

Additionally, the Project Team pursued other industry and external benchmarking partners. Industry sectors researched and pursued (based on their use of information technology for critical business functions) included:

- Other utility or combined utilities (electric, gas, wastewater, telecommunications, water)
- Municipal or regional government
- Finance, banking, insurance
- Medical, health care, pharmaceutical, manufacturing
- Professional service organizations (information technology, education, etc.)

The Project Team contacted and solicited 20-25 benchmarking partners through telephone interviews and screenings. Several of these potential partners were selected and completed the surveys. After results were analyzed and validated, three of these companies were selected for on-site case studies:

- Regional Municipality of Halton
- Electric, Gas and Water Utility
- Leading Manufacturing Company

Additionally, two participating utilities demonstrated leading practices and were ultimately selected for case studies: EPCOR and Seattle Public Utilities. The case studies are further described in Chapter 3 – Case Study Examples and Appendix C – Leading Practices Examples.

In addition to the on-site validated case studies, all participating utilities were asked to identify areas in which they thought they had unique or interesting practices that would be of interest to other utilities. Several of the utilities volunteered information about a variety of practices. Refer to Appendix C for more information about these contributed practices.

## SECURITY PRACTICE FINDINGS

With the increased emphasis being placed on security issues, an additional focus area for Security was created. An additional survey was developed based on the National Institute of Standards and Technology (NIST) Security Self Assessment Guide for IT Systems (Special Publication 800-26), supplemented by Sandia National Laboratories Draft Cyber Security Vulnerability Assessment. This supplemental survey included metrics, practices questions, and essay responses focused specifically on information technology security. Since security is an extremely sensitive area, four utilities chose not to participate in the survey.

Responses to the security survey were analyzed in conjunction with the other survey responses. Detailed information on the survey results is provided in Appendix B. A summary of key findings regarding security practices is presented below and in Figure 2.18. Again, the complete survey is found in Appendix A.

- Passwords and access practices are viewed as good, however, individual screening may be missing (survey questions H3, H7, H8 and H12)
- Physical access, protection, and backup is in place (survey questions H10, H11, H14, H15)
- Detection/access practices are not reviewed or monitored (survey questions H16, H17, H18, H19)
- Policies and human resources processes are in place, however, training may be lacking (survey questions H21, H28, H30, H33)
- Systems are initially secure but tracking/auditing may be missing (survey questions H22, H23, H24, H25)
- Business continuity plans and follow-through may be lacking (survey questions H35, H36, H37, H38)

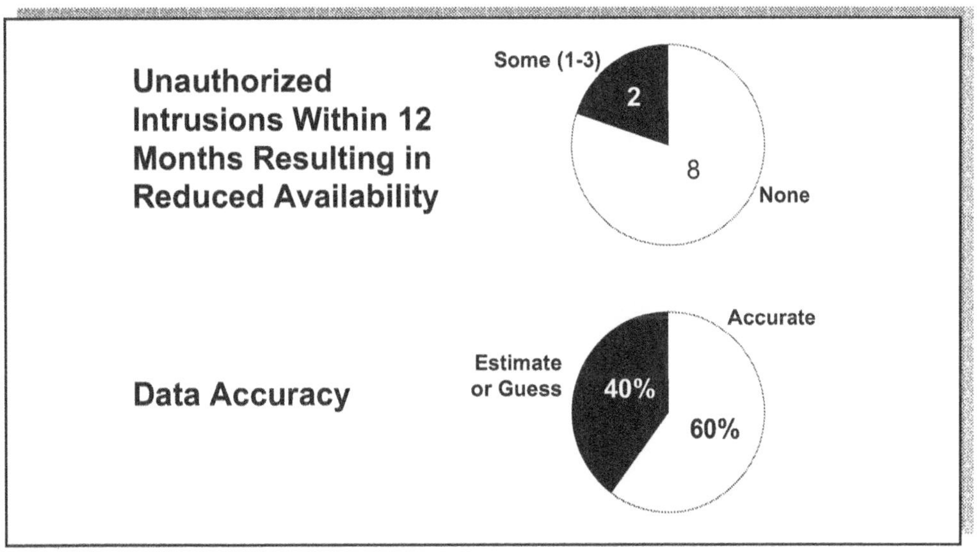


**Figure 2.18. Unauthorized intrusions are perceived to be minimal**

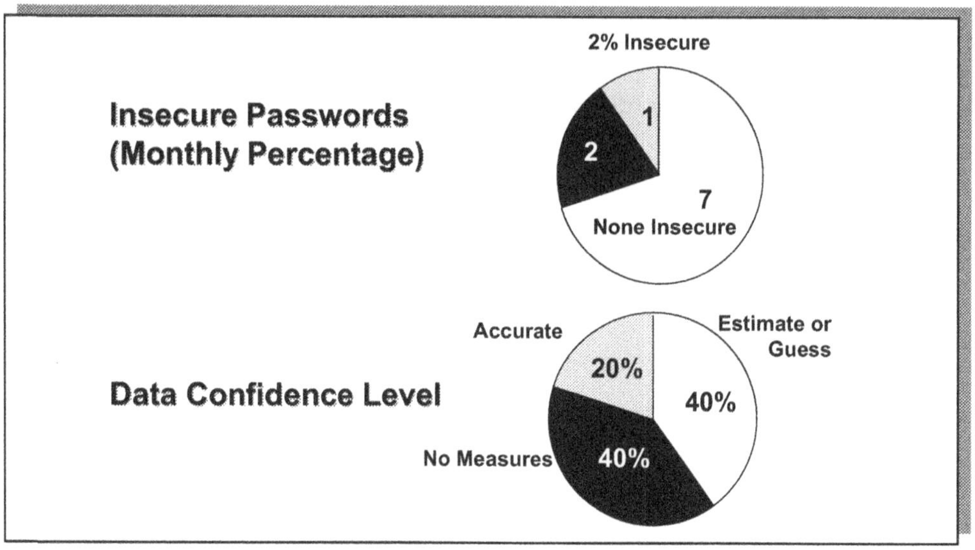


**Figure 2.19. Insecure passwords are low, however data confidence levels are also low**

Finally, respondents were asked to comment on their confidence level regarding their current IT security-related practices (Figure 2.19). Respondents indicated high confidence in security practices, with very knowledgeable and well-trained staff, and strong infrastructures.

# CHAPTER 3
# LEADING PRACTICES

In addition to the experiences of participating utilities, the project team analyzed experiences of organizations that actively demonstrated the leading practices, based upon their survey results, a phone interview, and on-site validation by one or two people from the project team. On-site visits included extensive interviews and first hand observations. Each case study focused on between three to five areas of leading IT practices for each company. The resulting case studies are described in this chapter. In three cases, the case study organizations were willing to share their identities; in two other cases the organizations preferred to remain anonymous, as follows:

- EPCOR
- Electric, Gas and Water Utility
- Leading Manufacturing Company
- Regional Municipality of Halton
- Seattle Public Utilities

The case studies addressed the following practices (Table 3.1):

**Table 3.1**
**Practices and case study organizations**

| Practice | Case Study Organization |
|---|---|
| Alignment with Strategic and Tactical Plan | Regional Municipality of Halton |
| Application Development and Enhancements | Electric, Gas and Water Utility, Regional Municipality of Halton |
| Application Testing | Seattle Public Utilities |
| Asset Tracking | Regional Municipality of Halton |
| Budgeting | Seattle Public Utilities |
| Customer Satisfaction | Electric, Gas and Water Utility, Leading Manufacturing Company |
| Data Definition (Models and Standards) | EPCOR |
| Data Integration | Leading Manufacturing Company |
| Governance and Policy | Leading Manufacturing Company |
| Help Desk | Seattle Public Utilities |
| IT Staff Development and Training | Electric, Gas and Water Utility |
| Network Management | Regional Municipality of Halton, EPCOR |
| Performance Management | Regional Municipality of Halton |
| Project Management | Leading Manufacturing Company, EPCOR |
| Roles and Responsibilities | Seattle Public Utilities |
| Service Level Management | EPCOR |
| Service Delivery Alternatives | EPCOR |
| Strategic Organizational Planning | EPCOR |
| Wireless | Seattle Public Utilities; Electric, Gas and Water Utility |

## CASE STUDY SURVEY ANALYSIS, OBSERVATIONS AND IMPLICATIONS

The results of the case study surveys were compared against the results of the other participating utilities. The following observations were noted when comparing the case study averages against the non-case study averages:

1. Case study organizations have a higher total annual revenue (for all services) per customer.
2. Case study organizations have a lower IT budget as a percentage of company operating budget.
3. Case study organizations have lower IT budgets as a percentage of revenue.
4. Case study organizations spend fewer IT dollars on a per IT staff basis.
5. Case study organizations have a higher percentage of internal IT staff.
6. Case study organizations have a higher training budget as a percentage of total IT budget.
7. Case study organizations support fewer PCs per IT staff.
8. Case study organizations show sufficient skills for most IT functions.

A number of possible implications could be drawn from these observations:

1. Case study organizations are not spending excessive amounts on IT relative to other organizations.
2. Case study organizations exhibit leading practices with fewer IT dollars per IT staff and have lower IT budgets as a percentage of revenue.
3. Case study IT organizations are adequately staffed.
4. Case study organizations exhibit leading practices by providing adequate (not excessive) levels of continuing education and training.
5. Case study organizations exhibit leading practices by providing sufficient (not excessive) staff per PC.

Three summary hypotheses can be drawn from these observations and implications:

1. Leading practices in IT can be achieved in an organization with sufficient but not excessive IT spending. The leading practices organizations spent less than 5 percent of their operating budget on all IT related costs.
2. When compared to other organizations, leading practice organizations may spend more for sufficient and skilled IT staff and reduce spending on other IT resources (such as hardware and software). The leading practice organization's IT staff averaged almost 10 percent of the total company full-time equivalents (FTEs).
3. Leading practice organization's demonstrated a clear focus on changing or redesigning their internal business practices using existing tools to gain efficiencies. This required that the IT staff provide more than just technical prowess, but also provide competencies and capabilities such as business understanding, effective communications, negotiation skills, meeting facilitation, and sound project management.

The following charts provide data to support the observations (Figures 3.1–3.8). The case studies are identified A through E. Further detail is provided in Appendix B.

- Observation: Total annual revenue (all services) per customer is higher for case study companies, on average, than non-case study companies.
- Implication: Case study companies could, in part, have higher revenue due to IT.

*Note that blank areas indicate that respondents did not provide data.*

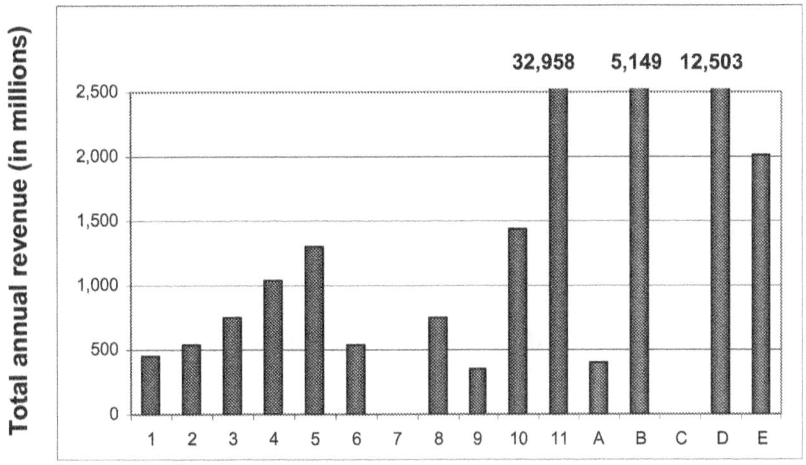


**Figure 3.1. Case study organizations have higher total annual revenue per customer**

- Observation: Case study companies on average have a higher percentage of internal IT staff than non-case study companies.
- Implication: Case study companies are adequately staffed.

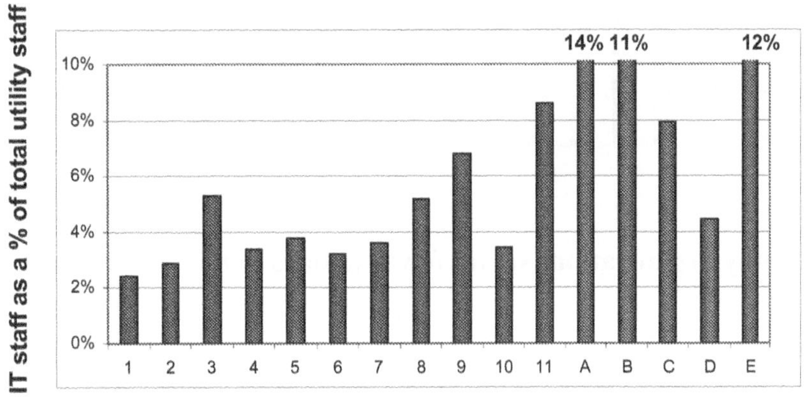


**Figure 3.2. Case study organizations have higher percentage of IT staff versus total utility staff**

- Observation: On average, case study companies have a lower IT budget as a percent of company operating budget.
- Implication: Case study companies are not spending excessive amounts on IT relative to other organizations.

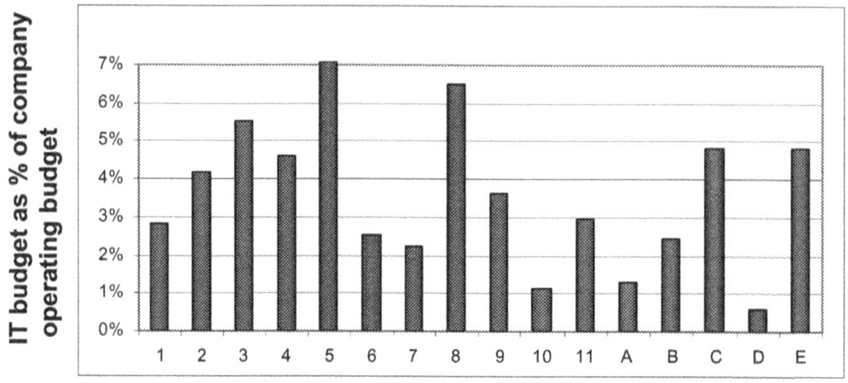


**Figure 3.3. Case study organizations do not carry the highest IT budgets**

- Observation: On average, case study companies spent fewer IT dollars on a per IT staff basis.
- Implication: Case study companies exhibit leading practices with fewer IT dollars per year, per IT staff.

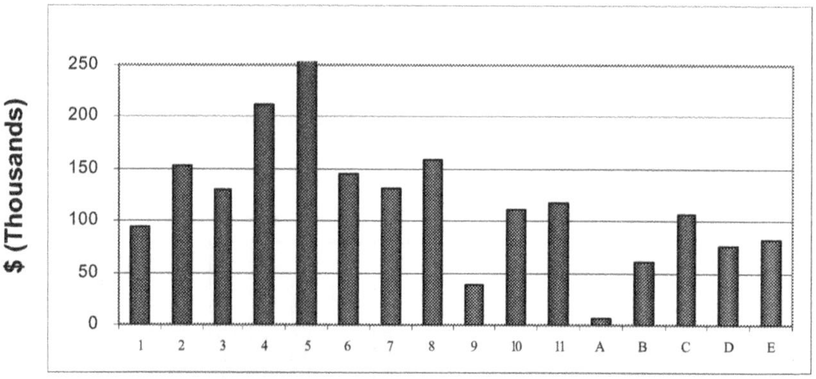


**Figure 3.4. Case study organizations spend fewer dollars per IT employee**

- Observation: On average, case study companies have lower IT budgets as a percentage of revenue compared to non-case study companies.
- Implication: Case study companies exhibit leading practices with lower IT budgets as a percentage of total revenue.

*Again, where there are blanks, respondents did not supply data.*

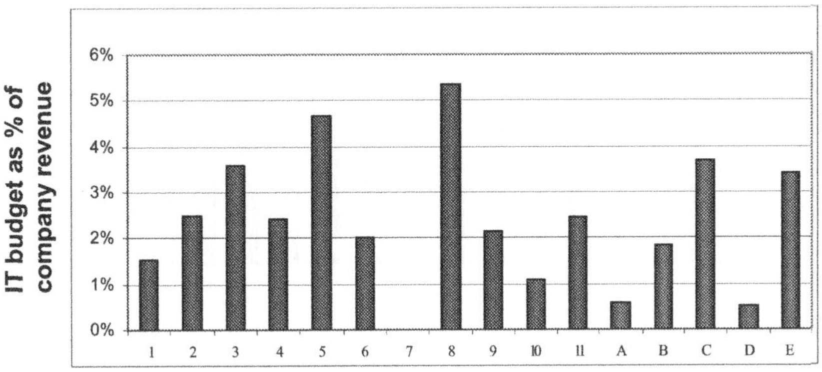


**Figure 3.5. Case study annual IT budget is lower as percent of revenue**

- Observation: On average, case study companies have a higher training budget as a percentage of total IT budget when compared to non-case study companies.
- Implication: Case study companies exhibit leading practices by providing adequate levels of continued education and training.

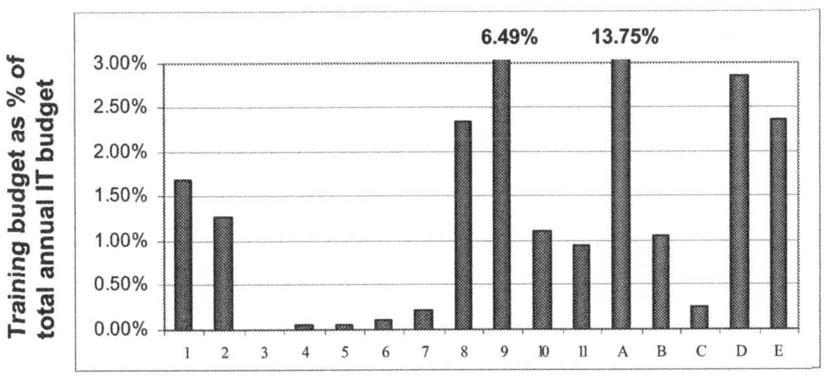


**Figure 3.6. Case study organizations percent of training budget is higher**

- Observation: Case study companies support fewer PCs per IT staff than non-case study companies. (Total PCs within the company/IT total staff.)
- Implication: Case study companies exhibit leading practices by providing sufficient staff per PC.

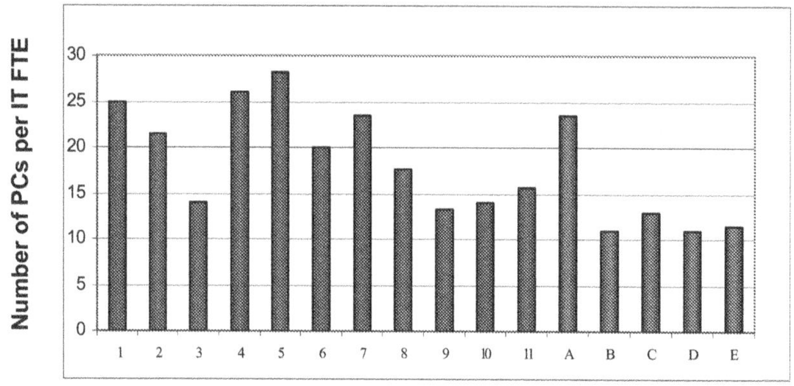


**Figure 3.7. Case study organizations support fewer PCs per IT staff**

- Observation: On average, case study companies (A through E) show that they have sufficient (balanced) skills for most IT functions.
- Implication: Case study companies exhibit leading practices by applying sufficient skills by IT function.

← Non-Case Study Companies (1–11) → ← Case Study Companies (A–E) →

| SH=Short Supply; B=Balanced; S=Surplus | 1 | | | 2 | | | 3 | | | 4 | | | 5 | | | 6 | | | 7 | | | 8 | | | 9 | | | 10 | | | 11 | | | A | | | B | | | C | | | D | | | E | | |
|---|---|---|---|---|---|---|---|---|---|---|---|---|---|---|---|---|---|---|---|---|---|---|---|---|---|---|---|---|---|---|---|---|---|---|---|---|---|---|---|---|---|---|---|---|---|---|---|---|
| | SH | B | S | SH | B | S | SH | B | S | SH | B | S | SH | B | S | SH | B | S | SH | B | S | SH | B | S | SH | B | S | SH | B | S | SH | B | S | SH | B | S | SH | B | S | SH | B | S | SH | B | S | SH | B | S |
| 1 General Management and Planning | ■ | | | | ■ | | | ■ | | ■ | | | | ■ | | | ■ | | ■ | | | | ■ | | | ■ | | ■ | | | | ■ | | ■ | | | | ■ | | ■ | | | | ■ | | | ■ | |
| 2 Business Application Analysis | | ■ | | ■ | | | | ■ | | ■ | | | ■ | | | ■ | | | ■ | | | ■ | | | | ■ | | ■ | | | | ■ | | ■ | | | | ■ | | ■ | | | ■ | | | ■ | | |
| 3 System Design, Development, Implementation | ■ | | | ■ | | | ■ | | | ■ | | | ■ | | | ■ | | | ■ | | | ■ | | | ■ | | | ■ | | | | ■ | | | ■ | | | ■ | | ■ | | | | ■ | | ■ | | |
| 4 Application and Data Maintenance/Support | | ■ | | | ■ | | | ■ | | ■ | | | ■ | | | ■ | | | | ■ | | | ■ | | | ■ | | ■ | | | | ■ | | | | ■ | | ■ | | | ■ | | ■ | | | ■ | | |
| 5 Infrastructure Maintenance/Support (Networks, Servers, PCs) | ■ | | | ■ | | | | ■ | | | ■ | | | ■ | | ■ | | | | ■ | | | ■ | | | ■ | | ■ | | | | ■ | | | ■ | | | ■ | | | ■ | | | ■ | | | ■ | |
| 6 Telecoms/Wireless Support | | ■ | | ■ | | | ■ | | | | ■ | | | | | | ■ | | ■ | | | ■ | | | ■ | | | | | | | ■ | | | ■ | | | ■ | | | ■ | | | ■ | | | ■ | |
| 7 Help Desk/Desktop Support | ■ | | | ■ | | | | ■ | | ■ | | | | ■ | | ■ | | | | ■ | | | ■ | | | ■ | | | ■ | | | | ■ | | ■ | | | ■ | | | ■ | | | ■ | | | ■ | |
| 8 Other (specify): Management | | | | | | | | | | | | | | | | | | | | | | | | | | | | | | | | | | | | | | | | | | | | ■ | | | | |
| 8 Other (specify): GIS | | | | | | | | | | | | | | | | | | | | | | | | | | | | | | | ■ | | | | | | | | | | | | | | | | | |
| 8 Other (specify): Management Analyst | | | | ■ | | | | | | | | | | | | | | | | | | | | | | | | | | | | | | | | | | | | | | | | | | | | |
| 8 Other (specify): Computer Room Operations | | | | | | | | | | | | | | | | | | | | | | | | | | | | | | | | | | | | | | ■ | | | | | | | | | | |

**Figure 3.8. Case study organizations have sufficient IT skills**

## CASE STUDY SURVEY DESCRIPTIONS

Descriptions of the case studies and specific leading practices are provided below.

### Case Study A: Regional Municipality of Halton

| | |
|---|---|
| Services Provided | – Regional Planning and Community Services<br>– Emergency Services (Ambulance, Police)<br>– Social Services and Health Programs<br>– School Boards<br>– Transportation<br>– Conservation Authorities<br>– Water Treatment/Distribution<br>– Wastewater Collection/Treatment<br>– Solid Waste Collection/Waste Transfer/Disposal/Recycling |
| Number of Customers:<br>Wholesale<br>Commercial<br>Residential | <br>0<br>5,500<br>94,800 |
| Annual Water Produced or Delivered (MG) | 11,631 |
| Annual Revenue<br>Water<br>Total | <br>$37,300,000<br>$40,539,000 |
| Annual Utility Operating Budget | $18,158,000 |
| Annual IT Operating Budget | $240,000 |
| Utility Staffing (FTEs) | 296 |
| IT Staff (FTEs) | 40 |

Halton demonstrated leading practices in:

- Performance Management (Figure 3.9)
- Application Development and Enhancements (Figure 3.10)
- Asset Tracking (Figure 3.12)
- Network Management (Figure 3.13)
- Alignment with Strategic and Tactical Plan (Figure 3.14)

These are further described below.

*Practice: Performance Management*

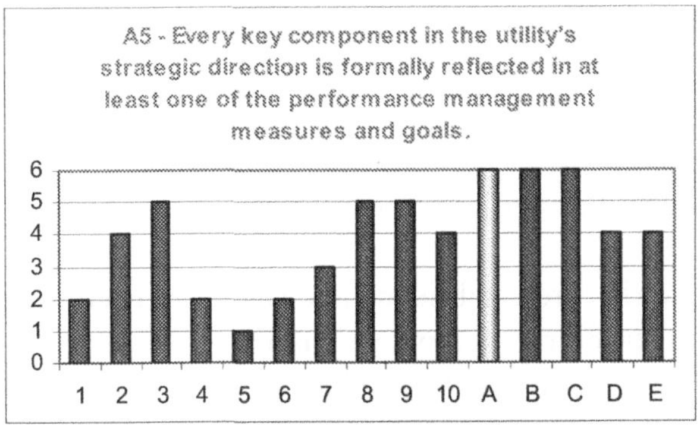


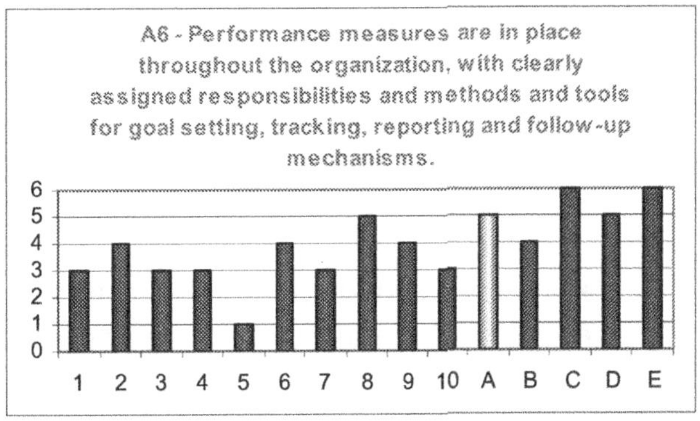


**Figure 3.9. Comparative performance management practices**

***Business Drivers:*** In 1995, the IT department was reorganized. This was driven in part by a large internal ERP initiative. The IT department, historically reporting through the finance department, now reported to the Vice President of Corporate Services.

Halton management believed that the IT department would be more accountable and successful by reporting to Corporate Services. Corporate Services had established a history of delivering products and services on time and within budget. Moving IT under Corporate Services direction automatically boosted the IT organization's image and probability for success with the planned ERP implementation. The organizational change was the catalyst for changing the performance management practices within the IT department.

***How Achieved:*** The ERP initiative was the catalyst for developing a long-term strategic and annual operating plan. The IT department now works with all department leaders to understand their long-term strategic objectives and to get an enterprise perspective of common synergies between departments. With a solid understanding of departmental long- and short-term project needs, the IT department plans their annual high level objectives and milestones. IT annual objectives are then broken down into written objectives for each member of the IT department. Each IT staff member has annual written development plans (objectives) for improving their skill set and competencies.

All department and individual objectives must be aligned with the corporate objectives. A standard format, supplied by the Human Resource department, is used for writing all performance appraisals and objectives.

***Benefits:*** The 1995 Enterprise Resource Planning (ERP) implementation was a huge leadership success for IT management. The project was delivered on time, within budget, with high customer satisfaction levels. This initiative showed corporate management that the IT department could take the corporate strategic plan of implementing an ERP system and manage all aspects of the project. As a result of this success, the IT department has continued to make numerous performance management changes with the IT department:

- Worked with Human Resources and senior management in developing a long- and short-term planning process for each department within IT

- Objectives were defined for each IT employee from departmental objectives
- Human Resources developed standard objective and performance appraisal forms and processes
- Written feedback (reviews) are completed annually by both the manager and the employee
- IT compensation is based on skill set and performance instead of years of service within non-union shops
- Managers review progress relative to objectives and coach employees throughout the year
- Annual and long-term IT performance metrics are established and regularly reviewed
- Grow the skill set of the department with structured competency plans
- Formal job descriptions supported a higher level of skilled employees

***Areas for Improvement:*** Establish Service Level Agreements (SLA) within all of the major business units. The SLA will tie back to IT departmental and individual employee objectives. The challenge is the "shock" of accountability associated with establishing SLAs. Some business units are reluctant to establish SLAs because they too would be held accountable when SLAs are not met.

Halton is planning to establish a planning or steering committee that oversees the IT function and its activities. This committee will include representation from the organization's senior management, users and the IT functions. The initial charter of the committee will be to establish governance and policies.

***Practice: Application Development and Enhancements***

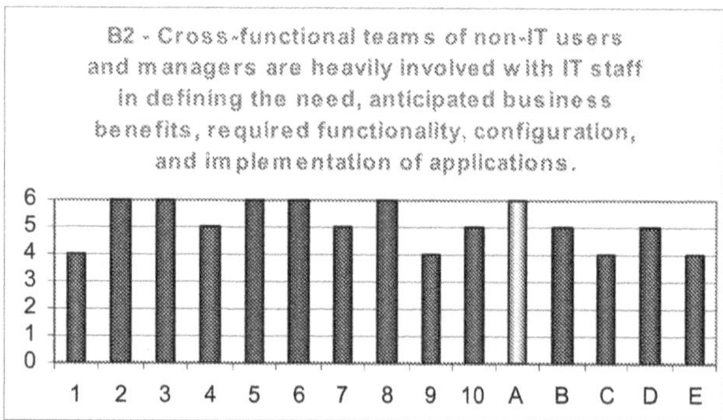


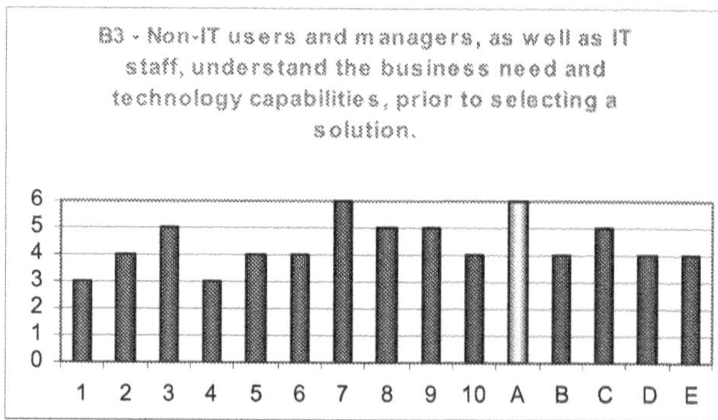


**Figure 3.10. Comparative application development and enhancements practices**

***Business Drivers:*** In 1995, the IT division developed a guide for developing business applications for the regional municipality of Halton. The application guideline, called The Halton Way, is intended to provide a concise, and understandable explanation of the approach, philosophy, and style for developing business applications. The Halton Way guideline, coupled with a major reorganization of the IT department and the planned implementation of ERP application were the key drivers for successfully developing and implementing enterprise applications.

***How Achieved:*** The Halton Way is intended as a recommended reference for anyone involved in developing business information systems for the Region of Halton. This includes analysts and project managers, external consultants working on behalf of Halton, and end user staff who undertake any significant application development effort that must be supported at some point by the corporation.

The Halton Way guideline for developing business applications has a four-prong approach:

**Section 1** outlines the Systems Development Approach. This section deals with the deliverables required at each stage of the Systems Development Life Cycle (SDLC). The systems development approach consists of discrete but related steps as outlined in the diagram below (Figure 3.11).

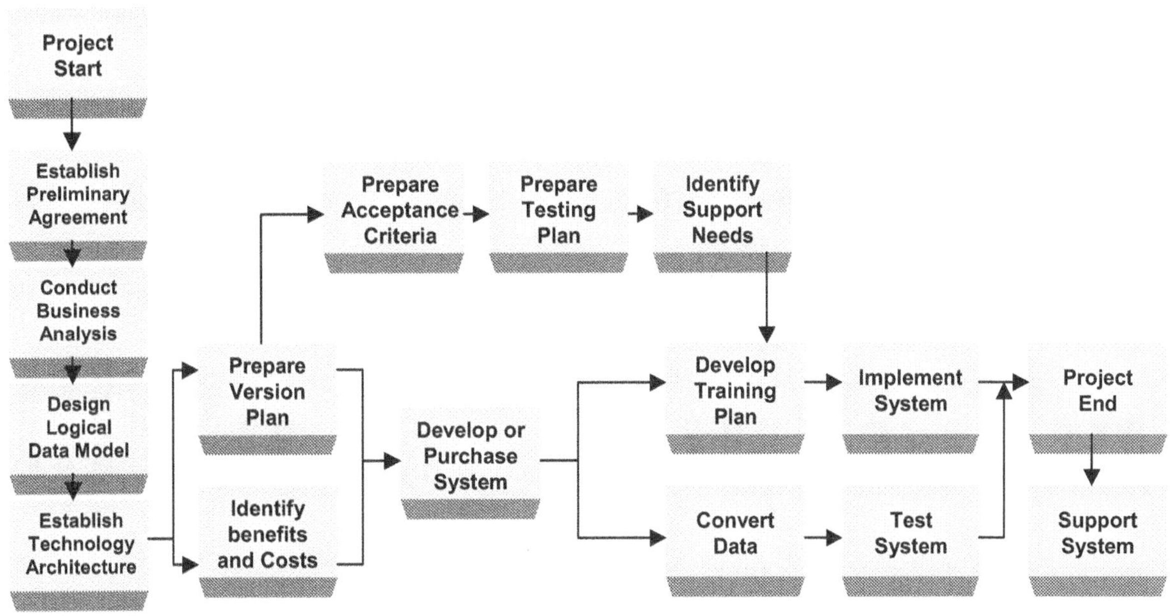


**Figure 3.11. Applications planning approach drills down to the lowest level**

**Section 2** outlines the Project Management Approach. It describes the project management tasks required for each project, such as project initiation and scope, project organization, project planning, status reporting, cost reporting, and change control management.

**Section 3** outlines the Resource Management Approach. It discusses resources planning, client resource requirements for the projects, project cost accounting, performance monitoring and outsourcing strategies.

**Section 4** outlines the Application Support Approach. It discusses end user and technical documentation, bug fix and program enhancements, performance reporting, help desk escalation process, and end user training.

***Benefits:*** The successful implementation of the ERP application was a "significant leadership success." The IT department has developed a reputation of meeting the needs of their customer and delivering projects on time and within budget. They are also credited with involving the end customer in all aspects of the development life cycle. The Halton Way guideline was so successful that all new projects now follow this process to receive funding. The success and stability of the ERP application has lowered the Total Cost of Ownership (TCO) for the various municipalities that use the application.

The guideline has reduced the scope creep associated with most major initiatives that do not have predefined project plans and deliverables. As a result of the guideline the frustration within the applications development team has subsided and rotation within the group has been reduced.

***Areas for Improvement:*** Educating the end customer on the value of thorough verification and testing processes and requiring their active participation to validate that the changes made have met their requirements. Business units who have actively participated with the validation and testing have been more satisfied with the end product.

Many municipalities or departments are reluctant to share their enterprise data. They understand the benefits of sharing data but politics disallow the practice. It also has been difficult to establish SLAs for enterprise applications in this environment.

***Practice: Asset Tracking***

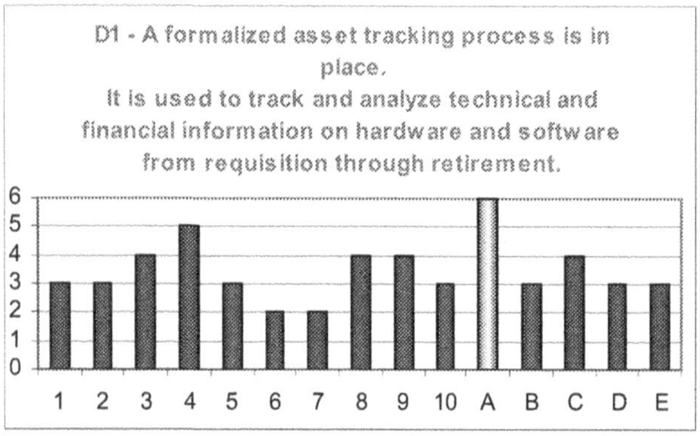


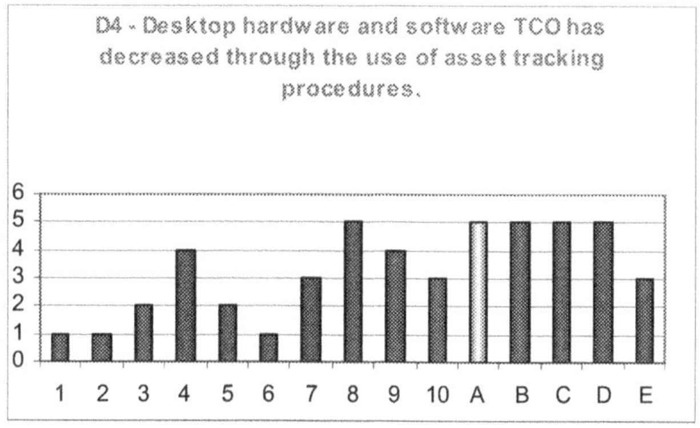


**Figure 3.12. Comparative asset tracking practices**

***Business Drivers:*** The need for an asset procurement and management practice within Halton was driven by the following factors:

- The proliferation of illegal and non-standard shrink-wrap applications installed by the employee on corporate PC equipment. Illegal and non-standard software installations increased IT maintenance and support costs, due to the additional time required by the IT department to resolve issues associated with the non-approved software installation.
- The lack of awareness of the total quantity of shrink-wrap applications purchased within the company. This made it difficult to negotiate maintenance contracts and leverage quantity purchases.
- The inabiltiy to uniformly and quickly distribute new software releases or upgrades to hundreds of PCs within the corporation.
- The ability for the IT department to accurately charge back departments for computer equipment used.

***How Achieved:*** The IT department assumed responsibility for procuring and tracking all computer and network infrastructure within the company. They consolidated all internal PC procurement and leveraged volume purchases for all computer equipment and shrink-wrap applications. They also standardized to one PC manufacturer and limited the desktop/laptop models from which to select.

Halton implemented Microsoft Systems Management Server (SMS) for managing many aspects of the company's PC hardware and software assets. SMS allows Halton to track PC equipment and related software by department, eliminating the charge back disputes from departmental managers. The automated software distribution within SMS has dramatically reduced the costs associated with the manual ("sneaker net") software upgrades. It also has intrinsic value by upgrading all employees at the same time.

***Benefits:*** The asset tracking system automatically generates monthly reports identifying equipment, by user, that is approaching the end of its lease term. Leased PC equipment can be returned to the leasing company as the lease becomes due. This eliminates late fees and additional monthly charges for past due equipment that previously could not be easily located. They can also quickly prove compliance with all software licensing and maintenance agreements.

When troubleshooting PC equipment, the IT department can quickly determine if the standard PC configuration has been modified or non standard software has been installed by the end user. This feature helps isolate potential issues and has proven to expedite the troubleshooting process.

***Areas for Improvement:*** With SMS the IT department can analyze the overall usage of departmental or individual hardware and software assets. A future goal is to fully leverage SMS to understand if assets are being effectively used. Example: If a laptop computer always remains connected to the company network (employee work area) for long periods of time, the employee may be better suited with a desktop computer. The laptop could then be redeployed to someone who requires a more mobile computer environment.

Halton would like to track field assets that are currently being manually tracked by regional departments. The IT department will be a key player with this initiative by assisting with data management of all assets. Developing an integration methodology will be the challenge for this project.

***Practice: Network Management***

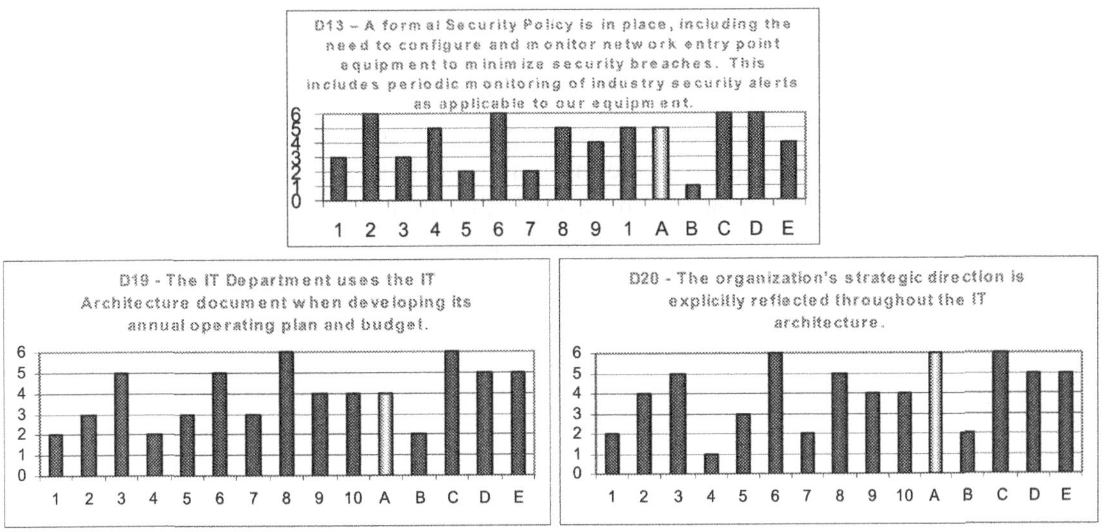


**Figure 3.13. Comparative network management practices**

***Business Drivers:*** By the mid 1990's, Halton had developed a number of network architectures. The networks were developed and maintained by numerous departments within Halton. This patchwork of unique network architectures and protocols made it difficult to seamlessly exchange data within the company. The planned implementation of a large SCADA system was the key business driver for moving the company to common network architecture.

***How Achieved:*** Halton senior management made the decision to centralize all LAN/WAN and PBX implementation and support duties within the company. The internal IT department was chartered to develop a migration play to transition the company to common network architecture.

The IT department developed a two-year transition plan for migrating the Halton LAN/WAN architecture to a TCP/IP backbone. A total of 40 different sites were transitioned to the new network architecture using 3COM network equipment. The larger facilities within Halton were designed with multi-point redundancy within the WAN. The redundancy ensures a higher probability of network uptime between facilities.

The IT department has also implemented a shareware software application (IP Century) for tracking and alerting IT staff when network and server capacity thresholds have been exceeded. They leverage a hardware sniffer (Network General and Fluke) for troubleshooting and resolving LAN/WAN issues.

When designing the new network architecture, the IT department designed in additional network capacity for their next generation phone system. Their long-term plan is to merge their voice, video, and data within their new LAN/WAN backbone. A pilot voice-over-IP (VoIP) solution using the Nortel product line has been implemented at one of the smaller Halton facilities. The pilot has been successful and there are plans to expand the VoIP solution.

***Benefits:*** The major benefit is having a common and reliable network backbone for exchanging data seamlessly. They also eliminated multiple Internet access points within the company to one single and secure corporate Internet access. Maintenance costs were reduced by 15 percent as a result of this initiative. A reduction in staff training, maintenance contracts, and FTE system configuration, maintenance, and monitoring costs are key contributors to the reduction.

The VoIP initiative has the potential for providing further cost reductions, as the solution is more widely deployed. There is also the potential for gradually eliminating the hard phone (the phone that sits on each office desk within the company) for a soft phone (located within the user's PC). Integration of phone, email and fax messages (unified messaging) is also possible.

***Areas for Improvement:*** Develop a comprehensive Disaster Recovery Plan (DRP) for corporate infrastructure. Halton is considering a DRP that includes an off-site facility to house their core network, servers, and applications. Multiple municipalities would share the cost of the DR site.

*Practice: Alignment with Strategic and Tactical Plan*

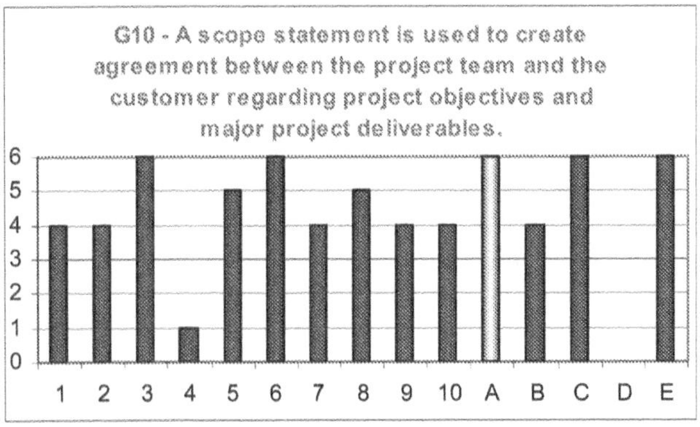


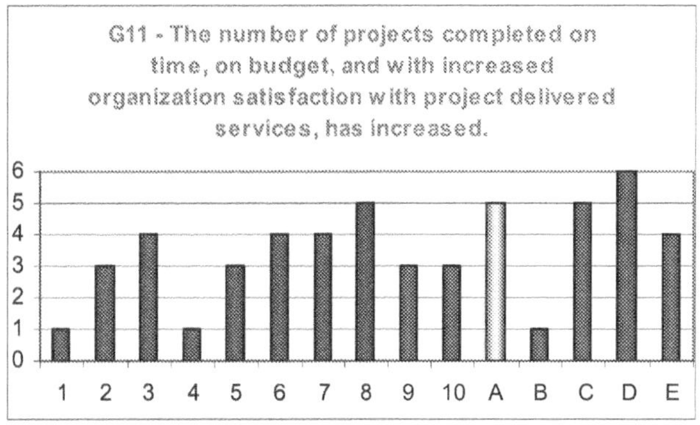


**Figure 3.14. Comparative alignment with strategic and tactical plan practices**

***Business Drivers:*** A Halton Region Technology strategy is developed every three years. The goals of the strategy are:

- To develop directions for technology that align with the region's business objectives of providing quality, cost-effective and accessible services to Halton residents.
- To identify the resources required to meet business demands.
- To develop a shared understanding of technology related directions, opportunities, and issues.

***How Achieved:*** Halton's philosophy is to build on the Region's existing technology strengths by applying new technology opportunities to address business challenges. The recommended strategic technology objectives are:

- To provide **robust technology infrastructure** that gives Regional staff access to corporate resources whenever and wherever they are required. This includes connecting the Region to its business partners and customer/clients through secure access to regional services.
- To implement **customer access technology** to enhance government services accessibility for Halton residents.
- To develop **services management information systems** required for effective management of Regional programs and to support performance management and benchmarking initiatives.
- To manage effectively the growing investment in technology and minimize the **total cost of ownership** of technology.
- To investigate and implement new technologies or new applications for existing technologies with **productivity enhancements** potentials.

To achieve the above strategic objectives and ensure alignment to the corporate business needs the IT department interviews all the corporate directors to learn their near- and long-term requirements. This information is distilled, researched, and common objectives are discussed in further detail in strategic planning meetings with the directors. Halton leverages external IT consulting services to assist with their strategic planning process. Consulting services are sought for technology management, utility regulations, and government agency management. When the strategic objectives have been defined, they're presented to the CAO and utility corporate counsel. Once the strategic objectives have been approved, an annual IT operating plan is developed.

***Benefits:*** Having the IT department focus on the five strategic objectives listed above, has played a key role in the overall success of the department and maintaining credibility with senior management. The strategic plan has also mandated the need for improving the professional skill set and competencies of the IT department. The IT department has made initial gains with formal education (e.g., MBAs, Project Management Professionals (PMP), Certified Management Consultants, etc.).

***Areas for Improvement:*** Building a tighter alignment with all business units and partners. This is expected to improve as the IT department continues to open communication channels with each of the business units and builds its credibility for delivering products that meet the needs of business. A tighter governance model with key success measures and alignment with Corporate Council is also desirable.

**Case Study B: Electric, Gas, and Water Utility**

| | |
|---|---|
| Services Provided | – Water Treatment/Distribution, Raw Water Transmission<br>– Wastewater Collection/Treatment<br>– Reclaimed Water Treatment/Irrigation Distribution<br>– Electric Generation/Distribution<br>– Natural Gas Distribution |
| Number of Customers: | |
| Wholesale | 0 |
| Commercial | 6,850 |
| Residential | 107,500 |
| Annual Water Produced or Delivered (MG) | 30,464 |
| Annual Revenue | |
| Water | $67,517,194 |
| Total | $588,796,856 |
| Annual Utility Operating Budget | |
| Water | $45,582,231 |
| Total | $443,606,172 |
| Annual IT Operating Budget | $10,777,469 |
| Utility Staffing (FTEs) | 1,620 |
| IT Staff (FTEs) | 179 |

The case study demonstrated leading practices in:

- Application Development (Figure 3.15)
- Customer Satisfaction (Figure 3.17)
- IT Staff Development and Training (Figure 3.18)
- Wireless (Figure 3.19)

These are highlighted below.

***Business Drivers (Overall):*** The utility is concerned about local politics moving to a privatization posture, to provide needed city revenue. Were that to happen, selling the utility would provide significant funds for the City. To prevent this from happening, the utility has an active program to turn ratepayers into active advocates for keeping the utility. The utility has adopted a "customer-owner" perspective throughout the entire organization. This is reflected in the utility's brand and image, as well as providing guidance for decisions.

The utility has recently reorganized to take advantage of their multi-utility environment (electric, gas, water, wastewater, reclaimed water). Prior to the reorganization, they were organized by line of business: electric, water, wastewater, etc. They are now organized by process: customer care, planning and finance, administrative services, operations, etc. Within operations are plant operations, field operations and control rooms. This reorganization has resulted in significantly fewer FTEs, as supporting services such as information technology were centralized and consolidated.

The multi-service environment has had a major impact on the utility's operational priorities, as electric deregulation and re-regulation movements have gained visibility. In 2002, the severe water drought brought attendant urgencies as mandatory water restrictions were imposed, along with a significant push to more effectively use existing water sources.

Management has created a vision for all employees with four major themes, related to their customer focus, dropping rates, and leveraging utility-wide resources wherever possible.

***Practice: Applications Development***

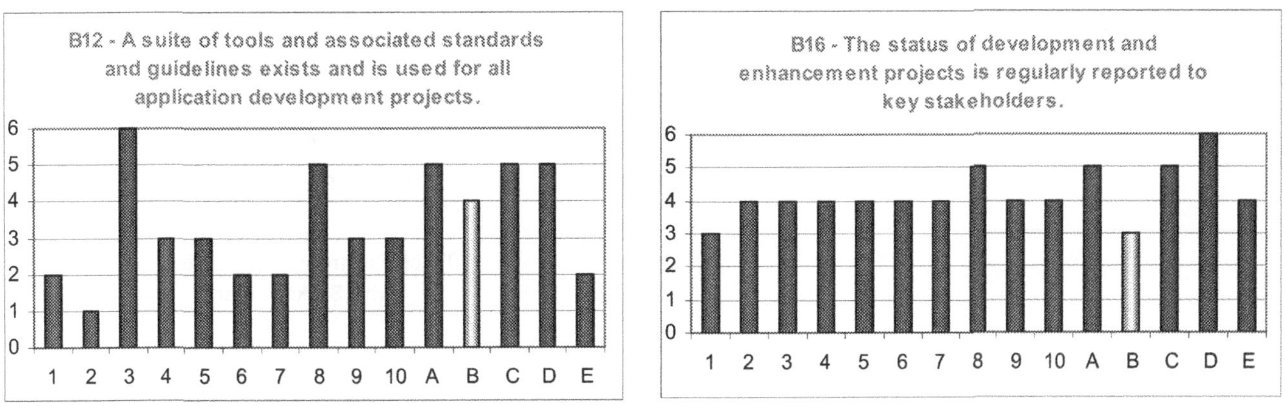


**Figure 3.15. Comparative applications development practices**

***Business Drivers:*** As part of the overall reorganization, several formerly separate IT organizations were merged into one unit. This highlighted that there were a number of duplicate applications. At the same time, business units were identifying the need for additional custom applications. However, there was no clear means of identifying what the new applications were supposed to do, how to compare the needs of one requestor to another, or even how many requests there were, and what kind of resources were being requested. The utility was not content to let the squeaky wheel get the application investments and developed a system for assessing requests.

***How Achieved:*** This new system allowed them to move from an unmanaged application portfolio to a managed portfolio – from over 400 applications to fewer than 100. Requests for application upgrades and enhancements go through a formal decision and scoping process (Figure 3.16). The process includes a formal request. All requests are assigned to a Business Services Analyst (BSA). The BSA develops a draft scope of work based upon the initial request. This is refined after interviewing the requester. At that point, an initial mapping of the findings and impacted datasets is developed and posted for review and comment. Upon a go/no go decision, the requestor is notified of the status. If a go decision is made, a work order and project plan are created so as to track the progress and keep stakeholders informed. Part of the decision-making process is the extent to which the request is related to other organizational initiatives underway, including Project Management, Customer Service, and Training.

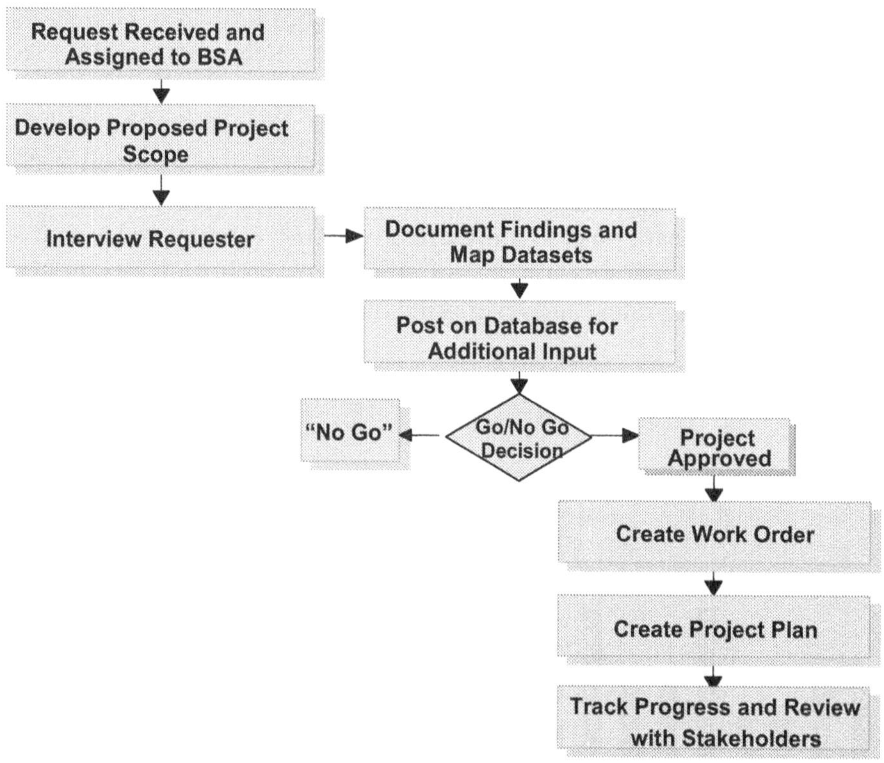


**Figure 3.16. Process flow for creation of new projects**

***Benefits:*** A number of benefits have been derived, including business unit customer satisfaction levels. IT can make clear commitments now, understanding the scope of work. Project plans are based on knowledge, not hope or guesswork. Overall business effectiveness has increased, since the business purpose and intended outcomes of enhancement requests are clear and identified up front. The results are tracked over time, lending additional credibility and highlighting areas needing more attention.

### *Practice: Customer Satisfaction*

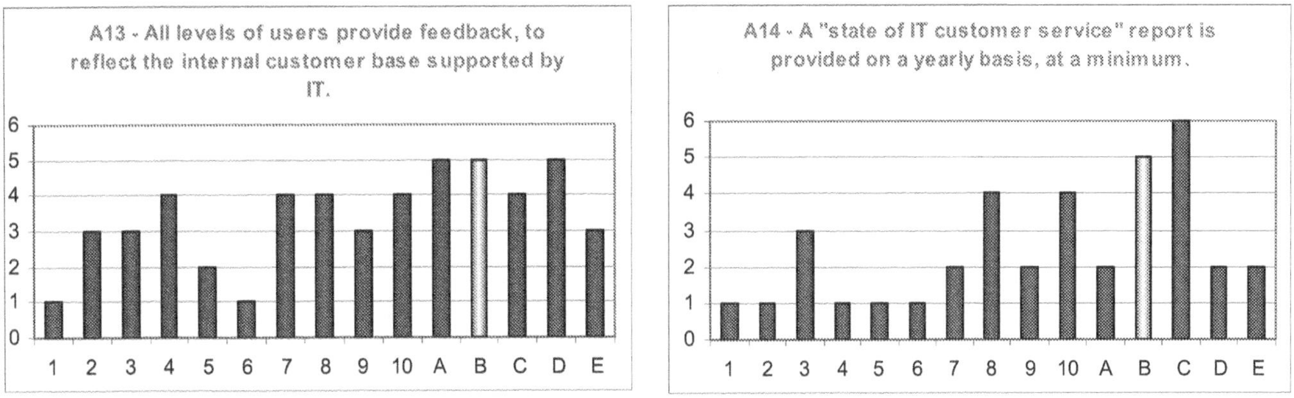


**Figure 3.17. Comparative customer satisfaction**

***Business Drivers:*** For years, this utility has conducted internal employee satisfaction surveys. It is part of their culture. The current emphasis on customer loyalty and retaining customers in a potentially deregulated electric environment has strengthened the core value of focusing on customer's needs, whether for "internal" or "external" customers.

***How Achieved - Internal:*** A yearly "employee climate" survey is taken. The response rate has been 90 percent or better over a six-year period. The utility manager heavily promotes the survey and all of the results are analyzed and reported at all organizational levels. The summary information is distributed to all employees and the results guide the utility. The feedback is integrated as part of managers' performance feedback. It has a direct impact on utility strategic and operational plans.

In addition, specific feedback regarding information technology employees is solicited on a transactional basis for the help desk and on projects when milestones are achieved.

All feedback is consolidated and reported down to the section level on a monthly basis. The results of the feedback are available utility-wide, and are posted on the intranet site.

***How Achieved - External:*** Ratepayer customer satisfaction and value have been actively tracked by the utility since 1995. The system used to measure this has been further advanced each year, with an outside consultant providing the ratings. The interplay between price, service, and relationship is closely examined. Because of their different interests, the utility has developed a different set of questions for commercial versus residential customers. The results of the external customer loyalty surveys directly impact the utility's strategic and tactical directions.

***Benefits:*** As a result of the ongoing focus on performance, the utility now has a data-driven way of getting overall feedback. They are able to see a macro view of trends, and understand whether performance is improving or not, and in what areas. They are able to use customer input as a driver for organizational changes. And, they are able to let their customers know how the utility is responding to customer input, in a highly quantifiable manner.

*Practice: IT Staff Development and Training*

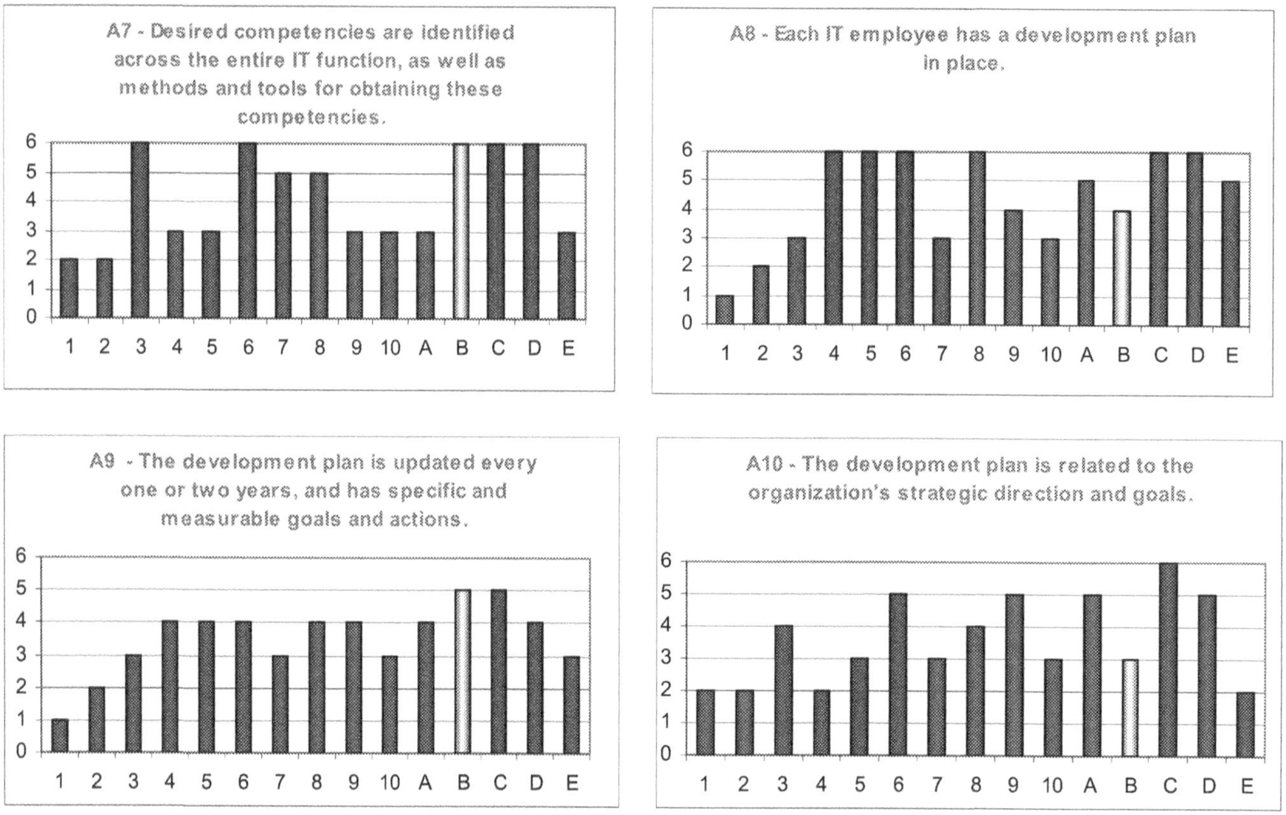


**Figure 3.18. Comparative IT staff development and training practices**

***Business Drivers:*** When the utility consolidated multiple IT sections, supervisors were in a situation where they did not know the skills of the people they managed. In addition, there was no clear view of available skills within the IT department overall. It was not possible to develop training plans that truly upgraded the department's capabilities. And, it was not clear if training was being provided in an area where there were already sufficient skills. The utility needed to gain an overall perspective of the current staff's skills and capabilities, and to link these to upcoming utility needs.

***How Achieved:*** An assessment form was developed which all employees in the IT department completed. The form focuses on current IT technical skills. Each employee rates their capability in various areas, with the ratings including no experience, some experience, working knowledge, and experienced.

Supervisors review the ratings with employees, to ensure consistent understanding. This provides the opportunity for new discussions and feedback regarding skills development. The skills are mapped against required skills for current and upcoming projects. Then a training plan is developed on the match of current versus required skills, and based on upcoming work.

***Benefits:*** A number of benefits have been seen, and training is now directly integrated into project planning – requirements of upcoming projects are mapped against employee growth objectives and current skills. Training is provided on a more timely basis, so that newly acquired skills are developed on actual projects. Training investments reflect business needs, not individual agendas. And finally, skills are leveraged across the organization more effectively because it is clear who has needed skills.

### *Practice: Wireless*

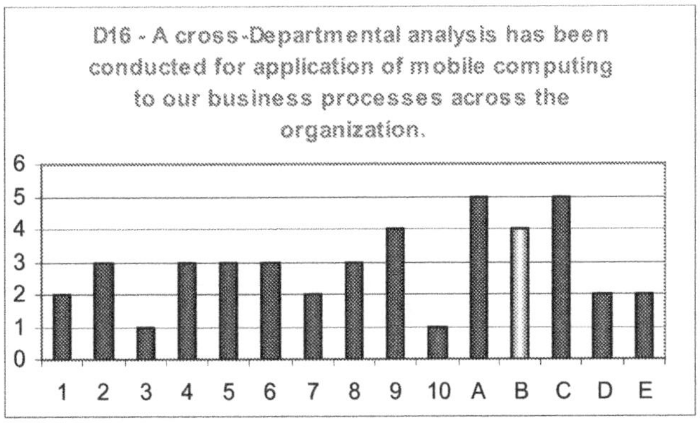


**Figure 3.19. Comparative wireless practices**

***Business Drivers:*** Deregulation/re-regulation issues in the electric business, and the local political climate have provided a strong focus on customer service for the past several years. This utility's focus on customer service included providing near real-time data to field service workers in the field.

***How Achieved:*** The utility has implemented an integrated wireless solution for customer service field workers. Throughout the process, end-users have been actively engaged in the hardware and software selection and implementation process. This solution provides service order information, inspection data, and schedule/route data in near real-time to all customer service people, regardless of location (in the office or in the field). The field workers log onto the system from home in the morning, receive their service orders, then go directly to the customer sites. They enter information at the time they get it in the field – there is no delay. Dispatch can send a field worker with the right skills, in the right location, to quickly address customer concerns when needed.

***Benefits:*** Immediate gains were seen due to the dramatic drop in paperwork. In addition, the cycle time for service request completions is greatly reduced – from several days to one day. And because data is entered one time, at the source of the data, there are far fewer errors. This has further reduced time spent correcting wrong data, thus freeing up additional time for other activities.

**Case Study C: Leading Manufacturing Company**

| | |
|---|---|
| Services Provided | – High Tech World-Wide Manufacturing<br>– Technology is coupled with manufactured products |
| Number of Customers: | Manufactured products benefit 2.5 million people |
| Annual Revenue | $6,410,800,000 |
| Annual Utility Operating Budget | $4,886,600,000 |
| Annual IT Operating Budget | $235,000,000 (~4% of revenue) |
| Utility Staffing (FTEs) | 27,731 |
| IT Staff (FTEs) | 2,200 |

The case study demonstrated leading practices in:

- Data Integration (Figure 3.20)
- Governance and Policy (Figure 3.23)
- Customer Satisfaction (Figure 3.25)
- Project Management (Figure 3.27)

These are highlighted below.

***Practice: Data Integration***

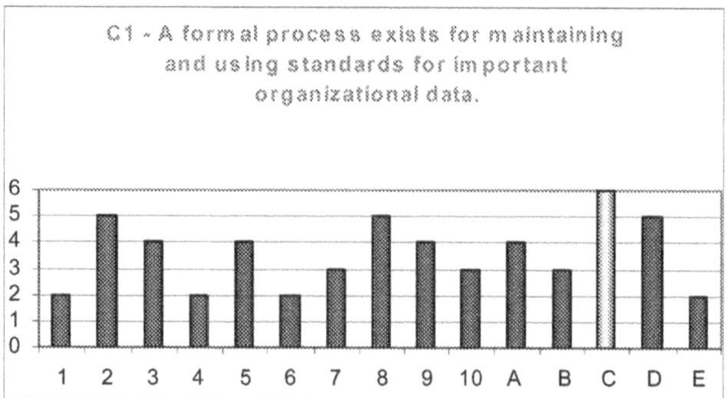


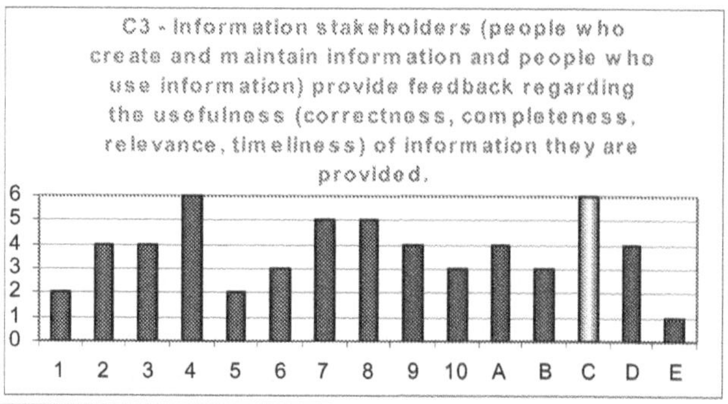


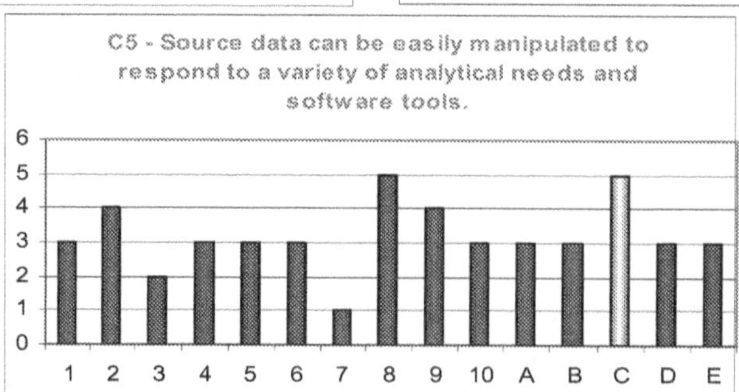


**Figure 3.20. Comparative data integration practices**

***Business Drivers:*** In the late 1990s, this manufacturing company had nine different major business units with nine unique database structures. There were a number of issues surrounding the various database methodologies and structures. Each business unit used a different catalog numbering scheme for raw material and finished goods. It was difficult to roll-up quarterly and fiscal year-end sales amongst all the divisions. Competitors were able to report financial performance within 2-3 days versus the 8-10 days required by this company. Costs associated with development and managing multiple business unit databases were excessive when compared to industry norms. Acquisition of additional companies also compounded the data management and integration issue.

A comprehensive data management strategy had many false starts within the company. There was a long-term history of each business unit having "unique" requirements that precluded them from developing a corporate data governance model. The board of directors, driven by the need to reduce the time required to report quarterly and fiscal year earnings, mandated that all business units would cooperate with building a comprehensive and integrated data management strategy (Figure 3.21).

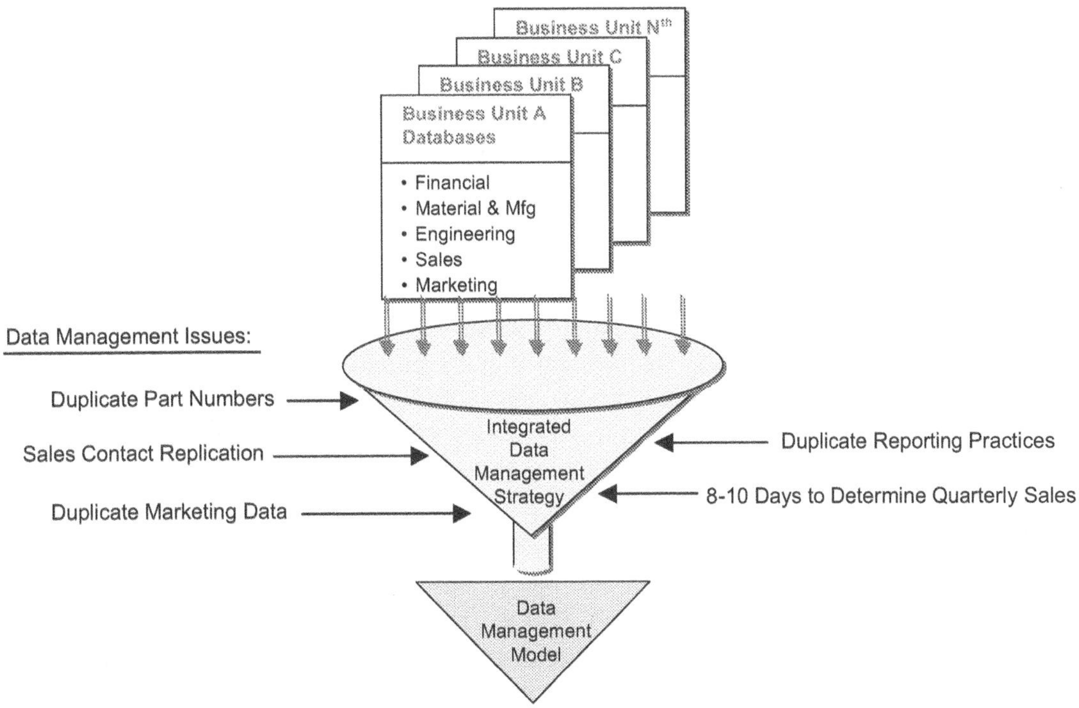


**Figure 3.21. Data management issues**

***How Achieved and Benefits:*** Executive management chartered a Corporate Data Management (CDM) swat team. This team, with representatives from each of the nine business units, was responsible for developing a single database development and maintenance team. The CDM team reported to the CIO of IT and the CIO made progress reports to executive management.

The CDM team first developed a governance model that all business units would abide by. Next, they developed a unified data architecture model. The unified data architecture model defined a process for the elimination of redundant data between business units. It also provided a methodology and transition process for common part numbers, engineering drawing numbers, invoicing, etc. Common database reports, revenue reporting, security policies and practices were also driven by the unified data architecture model (Figure 3.22).

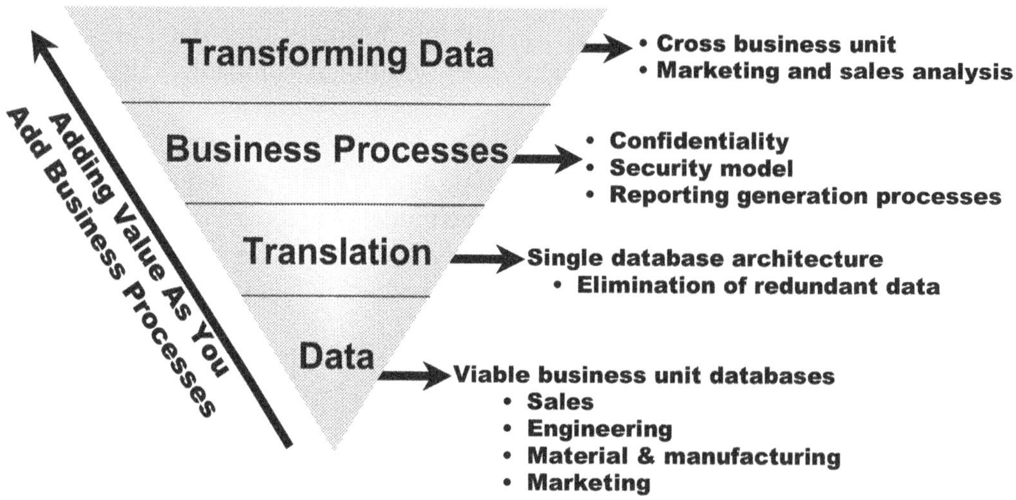


**Figure 3.22. Unified data architecture model**

When the unified data architecture model had been defined and approved by all levels of management, the databases from the nine business units where transitioned to the new data model. Ownership of the data was also symbolically transitioned to the corporate IT group. However, the new corporate database management and technical team is comprised of at least one member from each of the nine business units. Representation of each business unit with the newly formed database management and technical team was key to the initial success of this project.

***Benefits:*** This company can now provide financial reports within two to three days after the end of each quarter. The centralization of all enterprise databases and related management resulted in a reduction or reallocation of 18 FTEs. They also consolidated nine Oracle maintenance contracts resulting in an annual savings of more than $1 million. Data retention and replication processes are now consistent and follow leading practices.

*Practice: Governance and Policy*

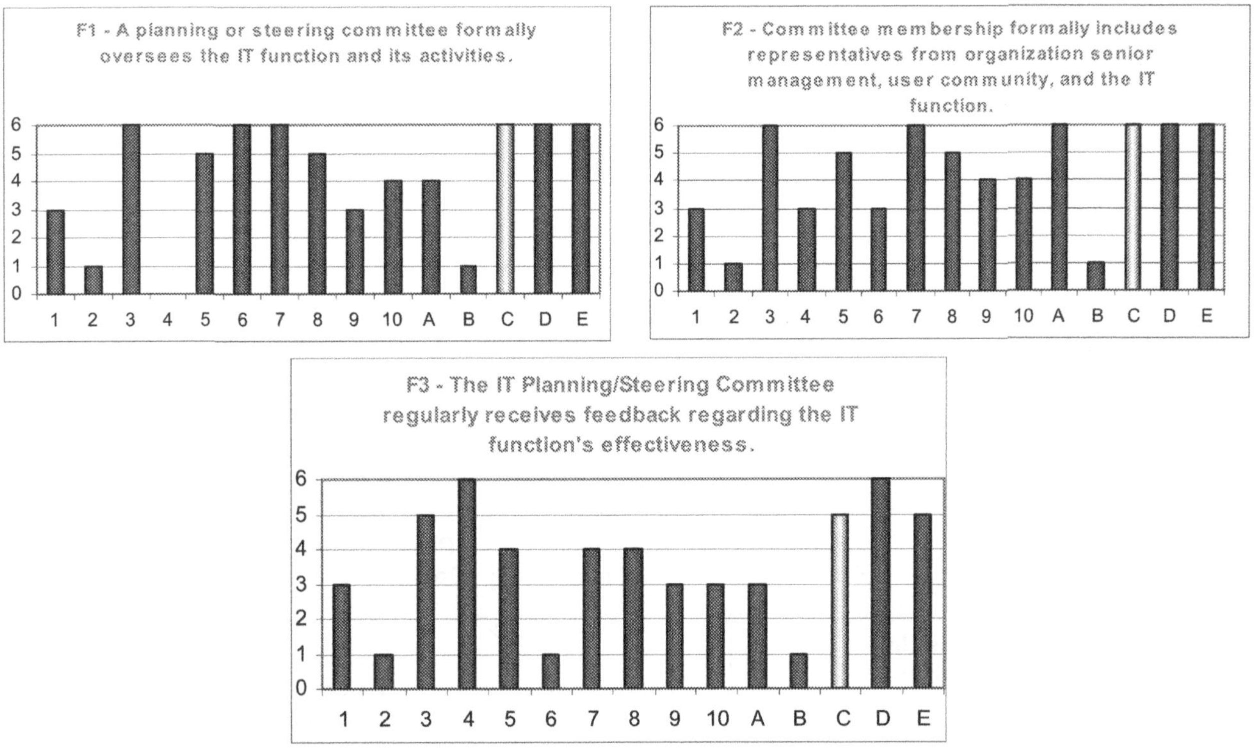


**Figure 3.23. Comparative governance and policy practices**

***Business Drivers:*** In addition to a corporate IT organization which provided core services throughout the entire company, this leading manufacturing company had nine business units each having their own IT departments. The long-standing justification for having siloed IT departments within each of the business units was due to their “unique” business requirements.

Each business unit felt that by having their own IT departments they were better positioned to quickly respond to their unique needs. This environment also created barriers between business units, negatively impacting the company. A key business driver for changing the governance within the IT organizations was driven by the lack of sales data integration between business units. Numerous attempts had been made to accomplish this over a ten-plus year period, however political barriers prevented IT department integration.

***How Achieved:*** Executive management created a CIO position and hired an external candidate from a Fortune 100 manufacturing company. All IT departments within each of the nine major business units were reorganized to report directly or by “heavy dotted line” to the new CIO.

The CIO identified the services offered by the new IT organization and presented an overview to executive management on the added value and strategic alignment that the new department could bring to the business at a cost savings of $2-3 million in the first year.

The CIO also established a Strategic Steering Committee to formally oversee the IT function and to ensure IT strategic and tactical objectives were in alignment with the corporate long-term goals. Business relationship managers who have responsibility for managing strategic and tactical planning within and across all business units were appointed to one (or more) sub committee within the Strategic Steering Committee (Figure 3.24). One member from each of the five sub committees sat on the Strategic Steering Committee. The members of this committee regularly met with the CIO to plan and review progress.

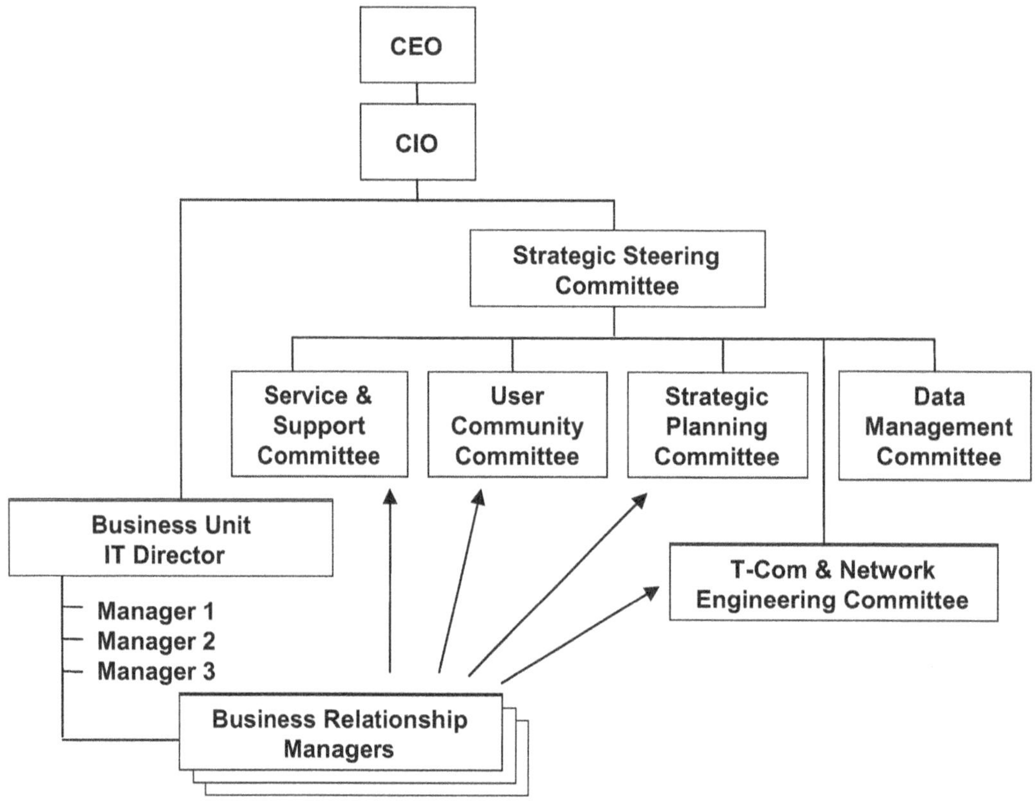


**Figure 3.24. Strategic steering committee**

***Benefits:*** Results are more quickly achieved and barriers are quickly broken down with the CIO reporting to the CEO. Over a 10 to 15 year period, this company had made many unsuccessful attempts to break down the political barriers within the various IT business units. In hindsight, they realized that having all of the IT business units reporting into a CIO position should have taken place years ago. They feel this is a leading practice.

*Practice: Customer Satisfaction*

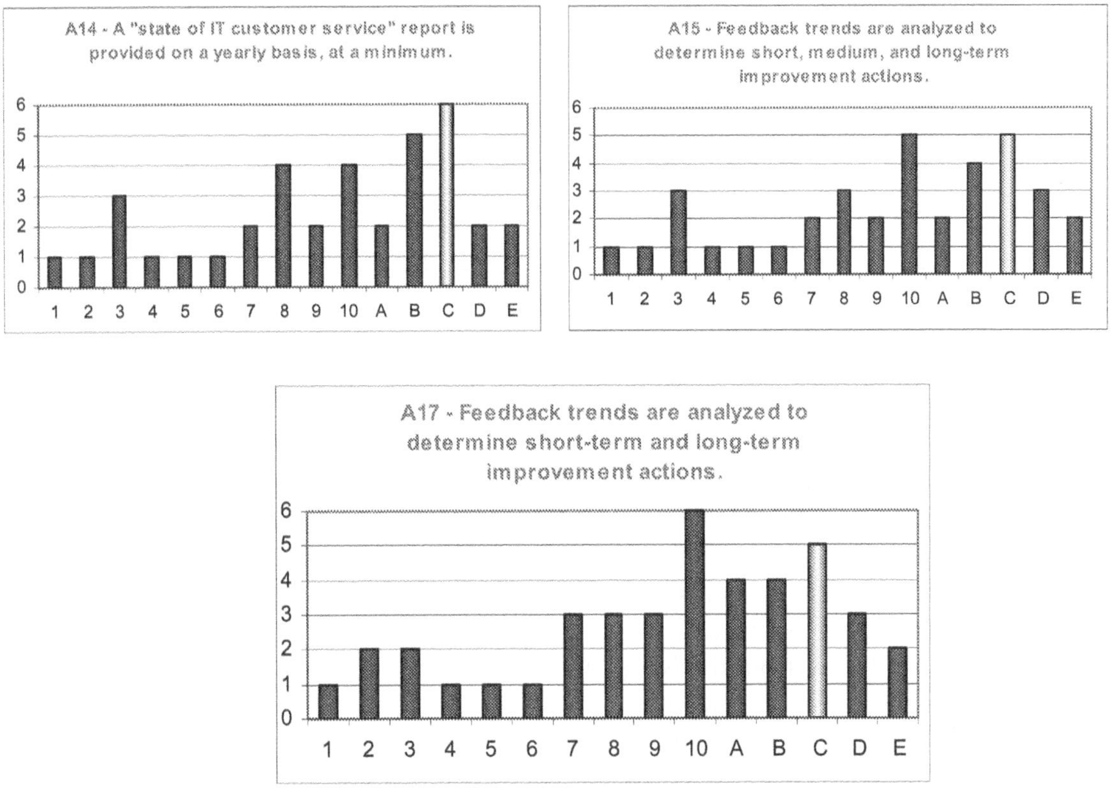


**Figure 3.25. Comparative customer satisfaction practices**

***Business Drivers:*** Each of the nine major business units within this corporation had their own IT organization. Internal customer issues, however, spanned more than one business unit. Customer satisfaction levels were below acceptable levels, in part, due to inconsistent issue tracking and common reporting tools for conducting issue-tracking analysis. Other issues included the need for flexible service level agreements per business unit, and a common set of metrics to measure end-customer satisfaction.

***How Achieved:*** The newly appointed CIO spearheaded an initiative outlining lost revenue to the corporation, and attributed it to delays with products reaching the market due to delayed or inadequate internal Service Level Agreements (SLA) and inconsistent practices for tracking and analyzing customer satisfaction.

As a result, a common issue tracking system and a baseline of core SLAs were implemented across all business units. These services were then centralized with the shared services IT department (Figure 3.26).

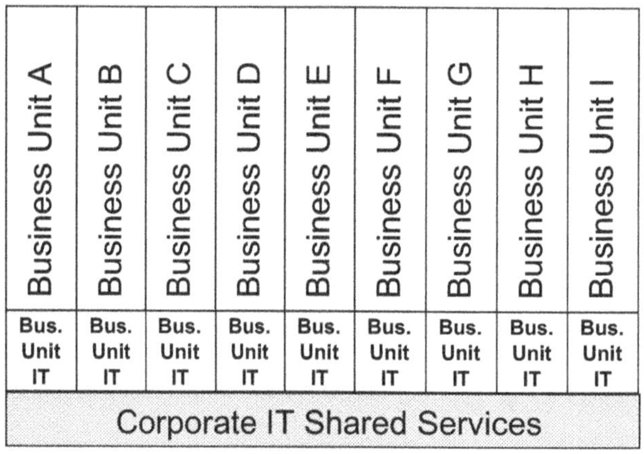


**Figure 3.26. Core SLAs provided by IT shared service to all business units**

Each business unit could pick from a baseline of SLAs that most suited their needs; it's analogous to an à la carte selection. Up to three additional SLAs could be established per business unit to address any unique customer requirements. All of the services selected could have three levels of support: gold, silver or bronze. The gold service provides the highest level of service at a premium cost. Business units with system or customer issues that are less critical typically select silver and bronze services.

Metrics were established from the SLAs. Customer satisfaction levels were measured by various survey methods. Short surveys were sent to 15 percent of all calls logged per month. The short survey was sent within 3 business days after a call is closed in the issue tracking system. The survey covered responsiveness to customer needs, knowledge of support staff, overall customer's satisfaction, etc. Semi-annual surveys were distributed to a wider audience. The survey was sent to an additional 15 percent of customers who placed a call with the help desk but were not covered in the previous short surveys. The semi-annual survey was also sent out to key stakeholders.

***Benefits:*** Time to resolve an issue dropped by more than 50 percent, as a result of implementing a common issue tracking system across all business units. Much of the time savings were attributed to the common features and benefits of a tier 1 issue tracking package. Costs were reduced on a per call basis by 15 percent over three years.

The à la carte selection worked well, allowing each business unit to select services that met their specific needs. The centralization of these services also created staffing efficiencies.

Business relationship managers (Figure 3.24, page 61) were responsible for meeting customer satisfaction levels within each of the business units. A percentage of their bonus was based on meeting the SLA. Feedback trends from customer survey data were analyzed by the business relationship managers to determine short, medium, and long-term improvement actions. Assigning customer satisfaction responsibility (active management) to the business relationship managers had a positive impact for meeting the agreed services within the SLAs.

*Practice: Project Management*

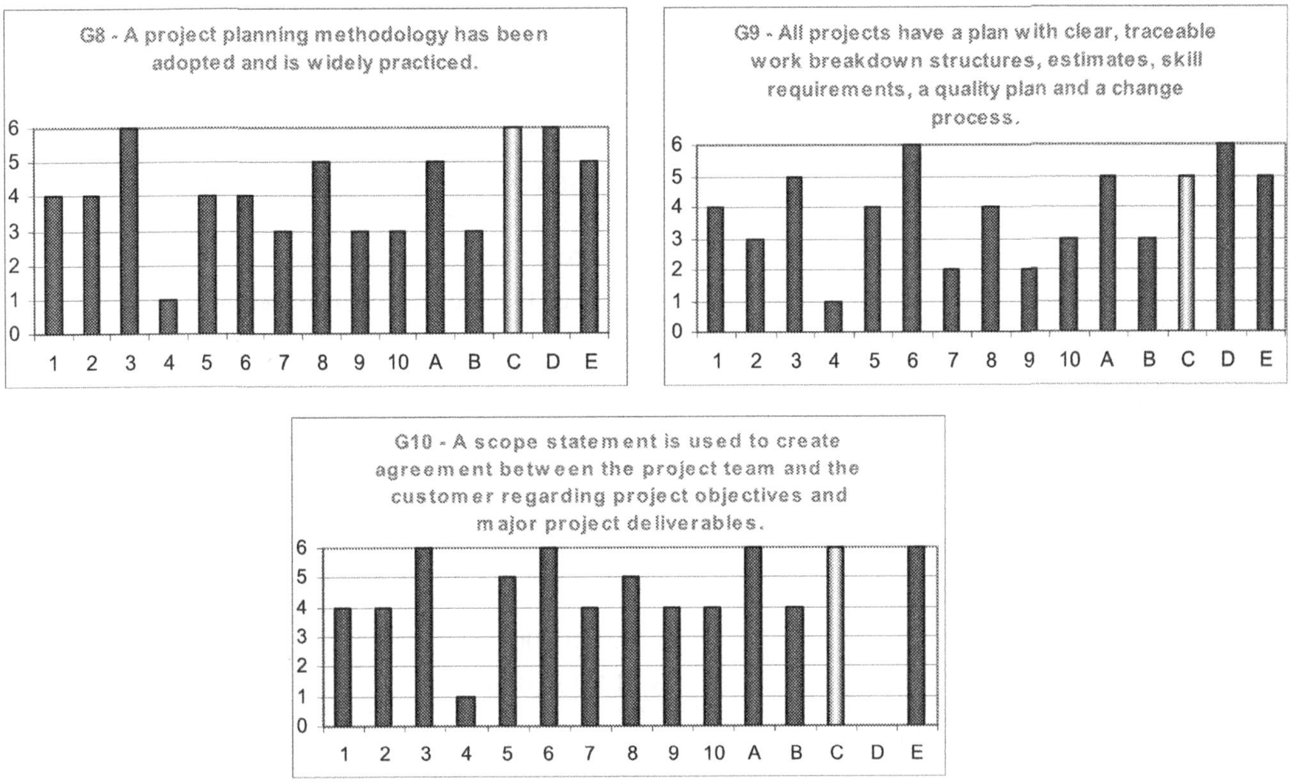


**Figure 3.27. Comparative project management practices**

***Business Drivers:*** Large projects spanned across multiple business units. Project management practices were inconsistent or non-existent between business units. Executive management believed product development cycle times would be reduced with a common project development methodology spanning all business units.

***How Achieved:*** A formal project development process methodology was used for all IT initiatives. All projects used Software Validation Development Life Cycle (SVDLC) methodology (proprietary document). The SVDLC document required the requesting party to provide an executive summary, overview of objectives, alignment to corporate objectives, measures of success, deliverables, resource requirements, risk management, quality control, communications management, life cycle costs, and integration management. Any project over \$20k used the SVDLC methodology. A senior IT manager sponsored each project. The IT executive committee reviewed all projects over \$20k. The CIO reviewed and approved all projects over \$50k and the corporate executive committee approved any initiative in excess of \$500k.

***Benefits:*** Business unit "pet projects" were immediately eliminated with the new project methodology and approval process. This process alone surprised management as to the increased availability of resources. Customer satisfaction has increased 25 percent with a 35 percent improvement for projects completing on time and within budget.

***Areas for Improvement:*** A process reengineering methodology is the next major initiative. Once this is developed, the first initiative will be reengineering the corporate

Enterprise Resource Planning (ERP) application. This will be a large project integrating 17 instances (revisions) of the EPR application across the business units.

**Case Study D: EPCOR**

| | |
|---|---|
| Services Provided | – Water Treatment/Distribution<br>– Wastewater Collection/Treatment<br>– Electric Generation/Distribution<br>– Merchant and Capital<br>– Power Development Corporation |
| Number of Customers: | |
| Wholesale | 7 |
| Commercial | 13,487 |
| Residential | 180,935 |
| Annual Water Produced or Delivered (MG) | 32,820 |
| Annual Revenue | |
| Water | $119,000,000 |
| Total | $3,714,000,000 |
| Annual Utility Operating Budget | |
| Water | $79,000,000 |
| Total | $3,230,000,000 |
| Annual IT Operating Budget | |
| Water | $1,229,399 |
| Total | $15,009,592 |
| Utility Staffing (FTEs) | 3,800 |
| IT Staff (FTEs) | 168 |

Also see Figure 3.28 for more information on EPCOR practices.

The case study demonstrated leading practices in:

- Network Management and Networks (Figure 3.29)
- Project Management (Figure 3.32)
- Data Definition (Models and Standards) (Figure 3.34)
- Service Delivery Alternatives (Figure 3.35)
- Service Level Management
- Strategic Organizational Planning (Figure 3.36)

These are highlighted below.

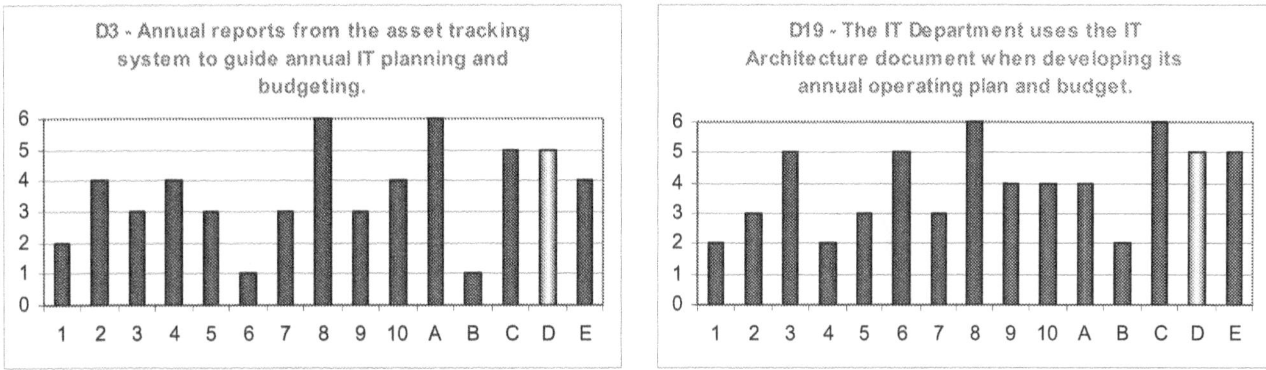


**Figure 3.28. EPCOR practices**

***Practice: Network Management and Networks***

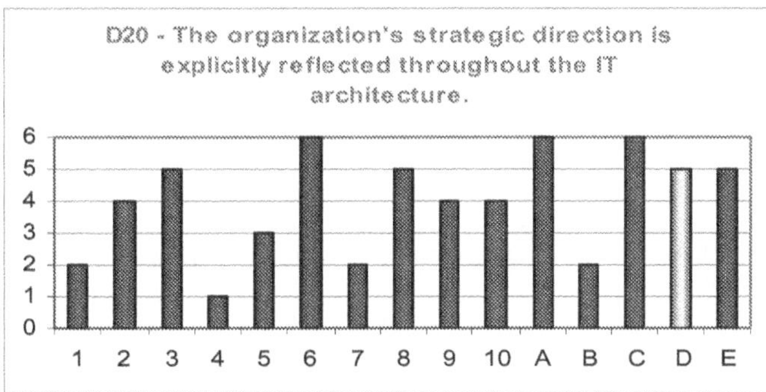


**Figure 3.29. Comparative network management and networks**

***Business Drivers:*** Computer networks and related managements tools are the backbone for moving information within EPCOR. The speed and reliability of the network and the agility of the network infrastructure team to quickly design and deploy LAN/WAN architecture for new acquisitions are key components to the business.

***How Achieved:*** The infrastructure division has developed a technology architecture roadmap that outlines their strategic planning cycles and technology strategy life cycle. It also includes tactical and long-term technology roadmaps for all major infrastructure/hardware components.

Technology roadmaps include:

- Client Device (PCs, PDAs, DVD, Printer/Fax/Copier Convergence, and Operating Systems)
- Tools (HP open view, Net, Oracle, Exchange, SMS)
- Networks (Gigabit Ethernet, Wireless PAN, VOIP, Wireless WAN, FIBER Channel, ATM)
- Server and Server Storage – (Unix, network servers, Tape, SAN, recoverability, clustering, disaster recovery)
- Security (Biometrics, advanced encryption standard, Kerberos, WEP Security, VPN, Firewall appliances, Data Encryption standard, Secure Socket Layer)

Each of the technology architectures is summarized on a maturity cycle roadmap. Figure 3.30 shows a hypothetical example of when the Pentium 4 – 2.xGhz PC will be deployed within its maturity life cycle. This one page summary graph easily identifies where EPCOR believes a specific technology is in terms of market maturity. EPCOR uses a maturity cycle roadmap as a method for communication to management and employees as to when they will possibly deploy new technology strategies within its maturity life cycle.

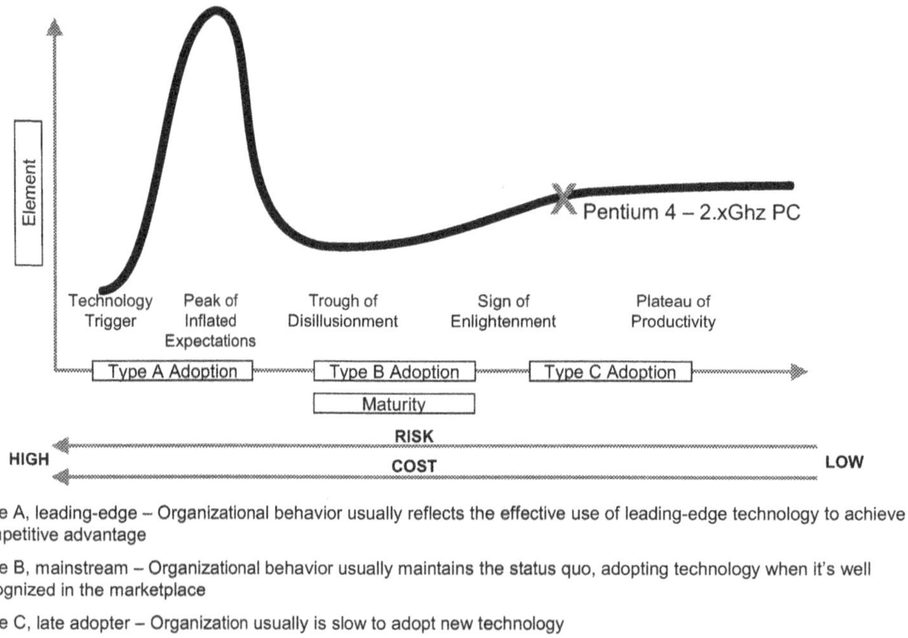


**Figure 3.30. Technology architecture roadmap guides decision-making and budget**

Figure 3.31 provides a finer level of granularity (time frame) as to when specific technology may be deployed within EPCOR.

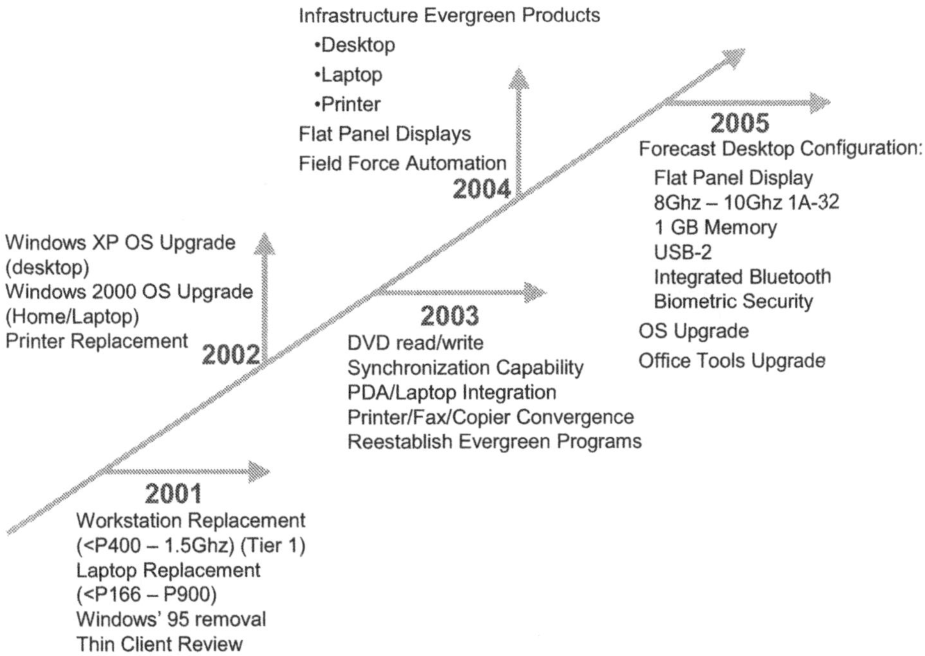


**Figure 3.31. Technology roadmap on client devices**

The technology architecture roadmap is reviewed every six months by an infrastructure steering committee. The steering committee, comprised of senior management, provides guidance to ensure the architecture being proposed is in alignment with corporate goals.

***Benefits:*** The technology deployed is usually mature in its life cycle, which mitigates the risk and ensures high probability of success. Capital costs for hardware equipment can also be lower because the product has been on the market for some time.

***Practice: Project Management***

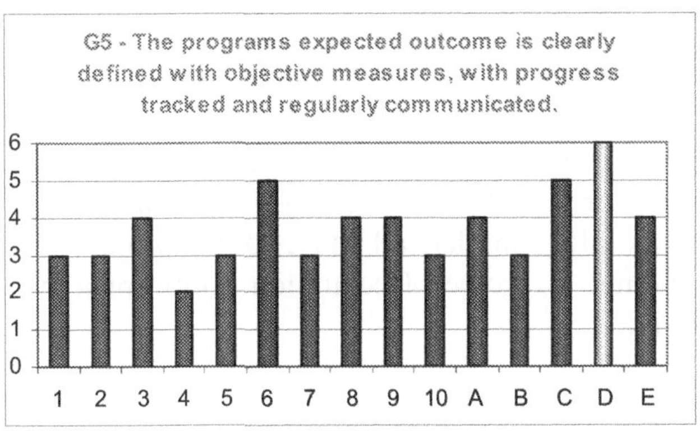


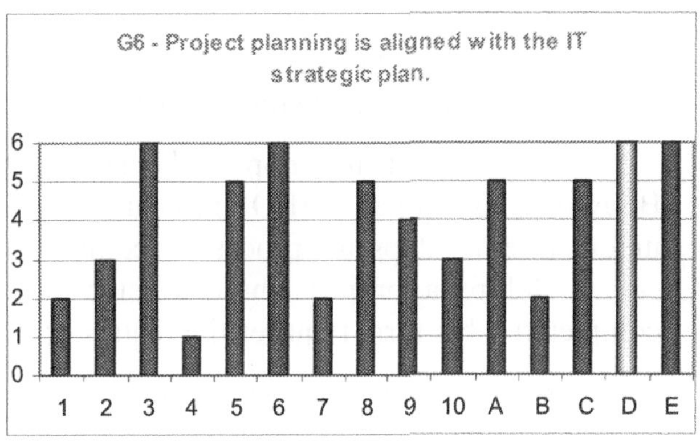


**Figure 3.32. Comparative project management practices (continued)**

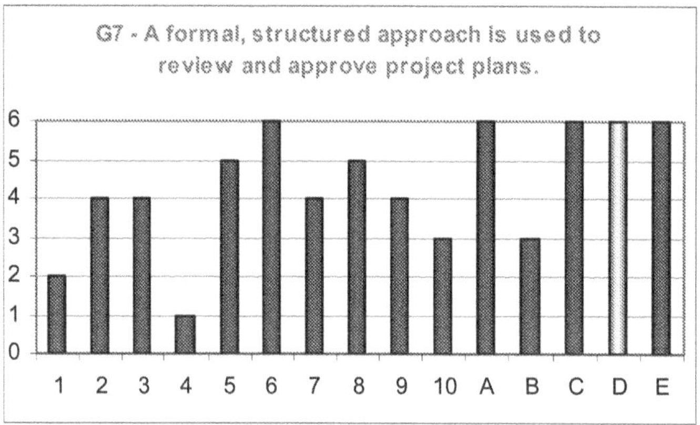


**Figure 3.32.** ***Continued***

***Business Drivers:*** Last year the EPCOR IT organization had over 140 projects in various stages of completion. With an annual budget of approximately $15m, there is a strong business need to understand the tools, services, and value the IT organization is delivering. Strong project management practices and standards are essential to ensure IT projects meet the near-term business needs and are in alignment with the long-term corporate strategy.

***How Achieved:*** The large volume of IT projects within EPCOR drives the need for a strong project management methodology and related practices. A Project Management Office (PMO) team was formed within IT and chartered to develop a structured project management practice. The team developed a comprehensive Project Management Manual (PMM). This manual has a comprehensive list of procedures, processes, and forms. Below is a brief summary of the PMM Table of Contents:

- Business Impact Report
- Business Requirements Report
- Service Level Agreements
- Business Design Report
- Detail Design Process
- User Acceptance Test Plans
- User Guides
- Project Actual Cost Form
- Post Implementation Review Process
- Package Evaluation Report

The PMO team adopted the Project Management Institute (PMI) Project Management Body of Knowledge (PMBOK) manual as a baseline for their PMM. The PMM also formalized unique internal steps and processes required to successfully complete all phases of the project.

Before a project phase receives any funding, a Project Declaration form must be completed. This form requires the requestor to provide:

- A summary of the project
- Objectives and linkage to a business plan
- Project strategy, assumptions, risks and potential issues
- A description of the measures of success with metrics
- High level deliverable with start and end dates

- Resource requirements and costs

The form is then reviewed by the IS workgroups (e.g., Operations, DBA, Applications, Security, etc.) who will be involved in completing the work if the project is approved. Each workgroup verifies that their involvement has been appropriately scoped. Signoff by each workgroup is desired but not required.

A senior level manager must sponsor each project and submit the completed Project Declaration form to the IS steering committee for review and approval. A detailed written project charter may also be required by the IS steering committee. The IS steering committee must approve all capital projects regardless of whether the project is planned or unplanned in the annual budget. Once approved, the first phase of the project may begin.

EPCOR uses a "gating" process to fund internal projects. The gating process holds the project manager to a specific dollar amount for each phase of the project. When a project manager completes a phase, or runs out of money in the current project phase, she/he must receive IT steering committee approval to move to the next phase. When projects are approved, they are reviewed monthly by the IT steering committee, which uses a bubble chart (Figure 3.33) to ensure projects are progressing as planned.

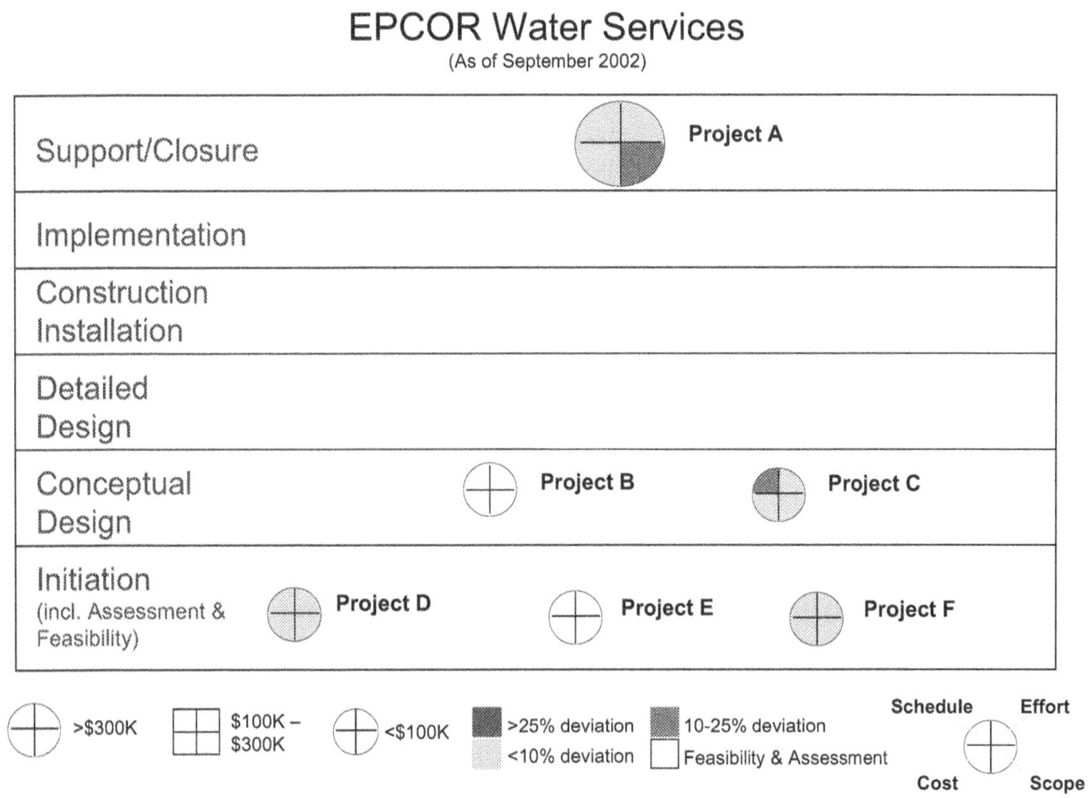


**Figure 3.33. Balanced scorecard bubble chart**

***Benefits:*** The project management process is flexible to accommodate both small and large projects ($50k-$10m). The established project management process ensures that projects are being prioritized and funded based on business strategy. The gating process holds all parties accountable and has significantly reduced project overruns and delays. It also allows the IT steering committee to control the number of projects being worked on simultaneously.

### *Practice: Data Definition, Models and Software*

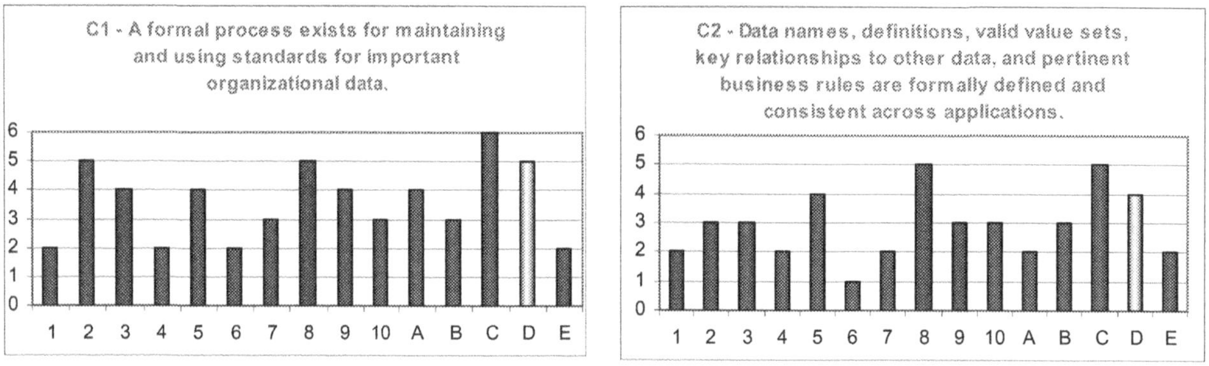


**Figure 3.34. Comparative data definition, models and software practices**

***Business Drivers:*** Improving corporate data management practices was driven by four main factors. Several databases within the company used duplicate baseline data. Managing and "synchronizing" this duplicate data provided little value to the business and was resource-intensive. There was a need to merge the "silos" of corporate data so that complex ad-hoc queries could be performed. Deregulation drove the need for a comprehensive data integration strategy that included an updated privacy, security, and data disaster recovery strategy.

***How Achieved:*** IT management created a data modeling position within the IT department. The near-term objective of this position was to create detailed standards for a data entity relationship model. Next, a common data architecture was developed to address the use of data modeling tools and practices. An entity relationship model was also developed to guide new database design, development and integration. Internal training programs were developed to train data and applications staff in leading practices using the Amadeus and Oracle Designer Suite.

They also leveraged modeling tools to reverse-engineer several applications and databases. The databases were cleaned up and integrated (e.g., Billing, GIS, Work Management, LIMS, Water Distribution).

***Benefits:*** The development of new data practices and tools substantially reduced the number of hours required to develop new, or modify existing, databases. Redundant data sets between business unit databases was reduced or eliminated. The use of data modeling uncovered gaps and inconsistencies in production databases.

*Practice: Service Delivery Alternatives*

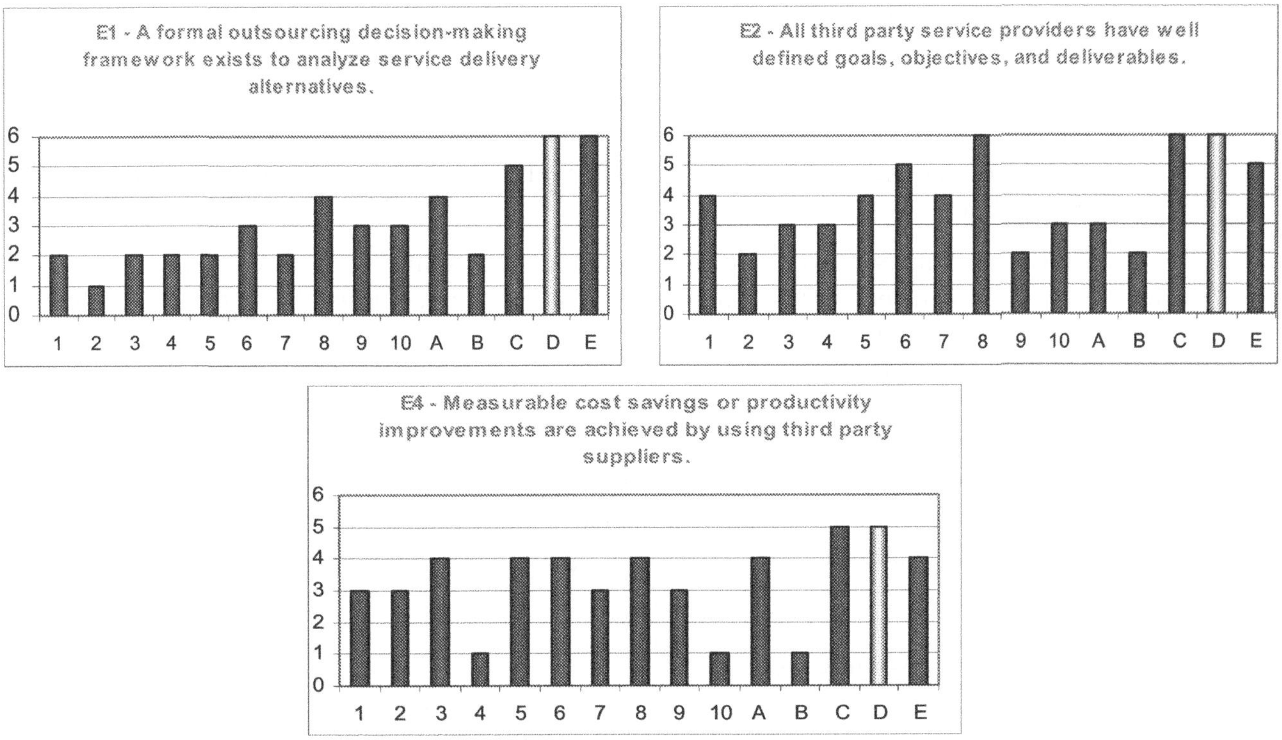


**Figure 3.35. Comparative service delivery alternatives practices**

***Business Drivers:*** EPCOR outsourced various IT services to multiple vendors for a number of years. The service level results from the outsourced vendors were mixed. The business driver was to change their outsourcing practice to reflect a higher level of service and accountability.

***How Achieved:*** The organization outsourced all network operations, desktop and server sourcing and support, and help desk services. To ensure effectiveness, it undertook a formal project to evaluate the business case for these practices. The project, conducted jointly with a tier-one consultant, completed a Total Cost of Ownership (TCO) analysis. The work revealed that while the current outsourcing alternative should be continued, a consolidation of vendors and contracts was necessary. A consolidation program was completed and new Service Level Agreements (SLA) were established.

The network operations, desktop and server sourcing and support, services were consolidated under one vendor. EPCOR IT management awarded the service contract to a smaller firm, as they were perceived as being more flexible in providing assistance that was not specifically written in the contract.

The vendor is compensated, in part, by the level of services provided. Below is an example of some of the services. Failure to meet the established metrics results in lost revenue for the vendor.

- Meantime to repair computer or server equipment

- Quantity of changes required after the release of a new systems configuration or software application. Example: After the vendor has deployed Exchange 2000, what additional server downtime is needed to resolve unplanned issues?
- Customer satisfaction on each help desk call
- Length of time a customer is holding on the phone before the call is answered by the Help Desk
- Many of the network related metrics within the SLA use data provided by HP OpenView

EPCOR continuously looks to improve their internal IT services. They focus on developing or improving processes that would have the greatest impact on the end customer. This is accomplished, in part, by conducting monthly meetings that focus on near- and long-term continuous process improvements.

***Benefits:*** By reducing the number of outsourced vendors, EPCOR has reduced the number of hours required by internal staff for managing these services. They also use outside consulting firms to ensure that their outsourcing practices, customer satisfaction, and related costs are in alignment with other leading practice companies. A benchmark analysis of their outsourced services indicates that they provide a higher level of service to their customers at 25 percent lower cost.

***Practice: Service Level Management***

***Business Drivers:*** As EPCOR moved toward outsourcing some of their IT services, there was a need for effectively managing SLAs for both internal and outsourced services.

***How Achieved:*** During the process of developing new SLAs for IT Operations services, EPCOR adopted a new methodology from the Infrastructure Technology Information Library (ITIL). The ITIL is a series of documents that aid in the implementation of a framework for IT Service Management (ITSM). ITSM is divided into two main areas, Service Support and Service Delivery.

Service Support focuses on five disciplines: Configuration Management, Incident Management, Problem Management, Change Management, Service/Help Desk, and Release Management. Service Delivery is the management of IT services, and involves a number of management practices to ensure that IT services are provided as agreed between the service provider and the customer.

With the ITSM documentation and assistance from an outside consulting firm, EPCOR established new practices for implementation and management of their IT services. The SLAs were developed to include more flexible scopes, performance metrics, incentives, incident management with root cause analysis, and continuous improvement metrics.

***Benefits:*** The ITSM methodology helps foster a partnership between the service provider and the customer. It has led to more cooperation between parties, which allows for more flexible SLAs. EPCOR has also achieved reductions in system outages and has defined more customer focused Help Desk metrics.

*Practice: Strategic Organizational Planning*

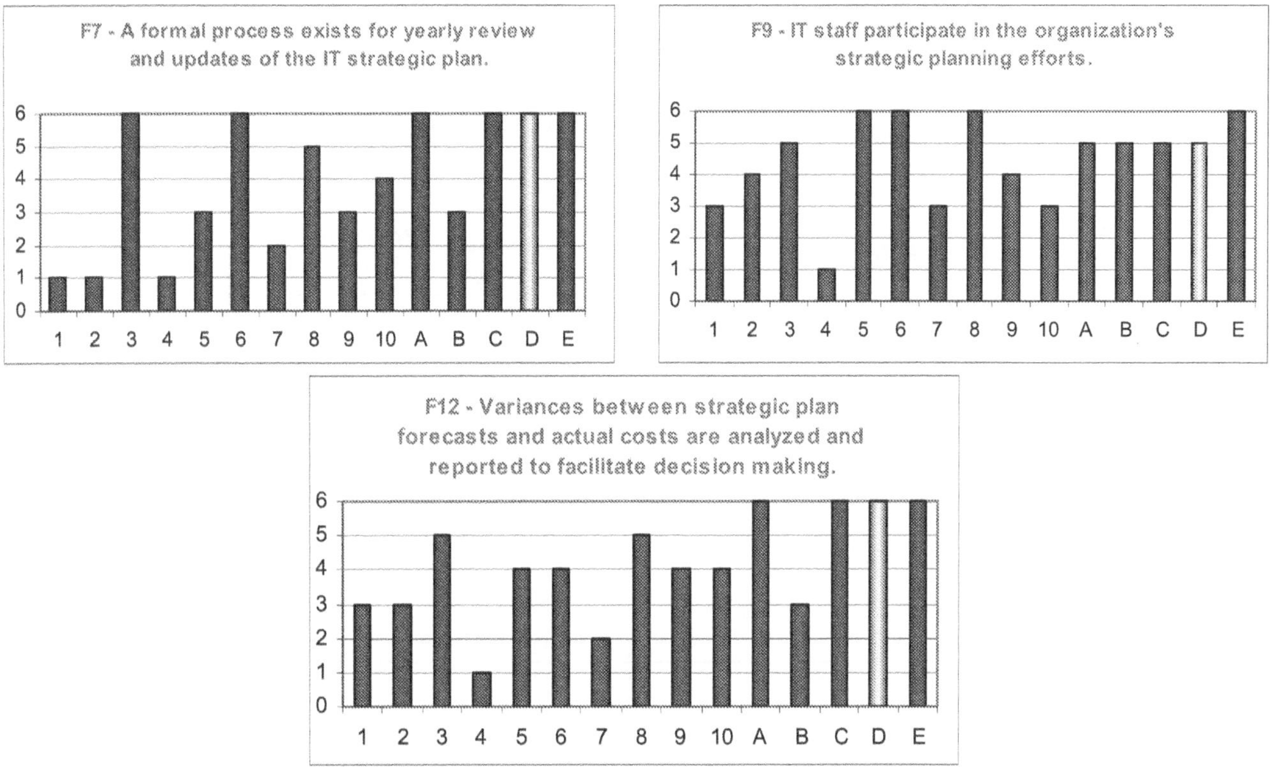


**Figure 3.36. Comparative strategic organizational planning practices**

***Business Drivers:*** An IT strategic plan reflects an organization's IT vision, mission, and goals. Strategic planning is a management tool used to focus an organization's energy to ensure that members of the organization are working toward the same goals and to assess and adjust the organization's direction in response to a changing environment. EPCOR has created an IT steering committee consisting of business unit and IT savvy management. The goals of this team are to ensure that IT and the business units collaborate to successfully create, execute, and manage IT services for the individual business units.

***How Achieved:*** At an enterprise level, an information services strategic plan is produced to establish a medium- to long-term broad perspective and analysis of the organization's information and information technology requirements. At a business unit level, EPCOR has implemented an IS steering committee. The steering committee focuses on the business needs and guides the IS organization to meet those needs. The group consists of senior IS and computer literate business unit staff representing each area. This team drives synergy within the business units that leads to economy of scale.

The steering committee meets once a month and has a set agenda. The agenda sets expectations as to what is expected from each team member and focuses on key initiatives or topics that are aligned with the corporate strategy. Below is a summary of the set agenda:

1. Review active projects – The committee reviews a one page summary of all major IS projects. The summary outlines which of six phases (initiation, conceptual design, detailed design, construction installation, implementation, and support/closure) a project is in. IT also identifies any deviations to schedule, effort, cost, and scope (Figure 3.37).

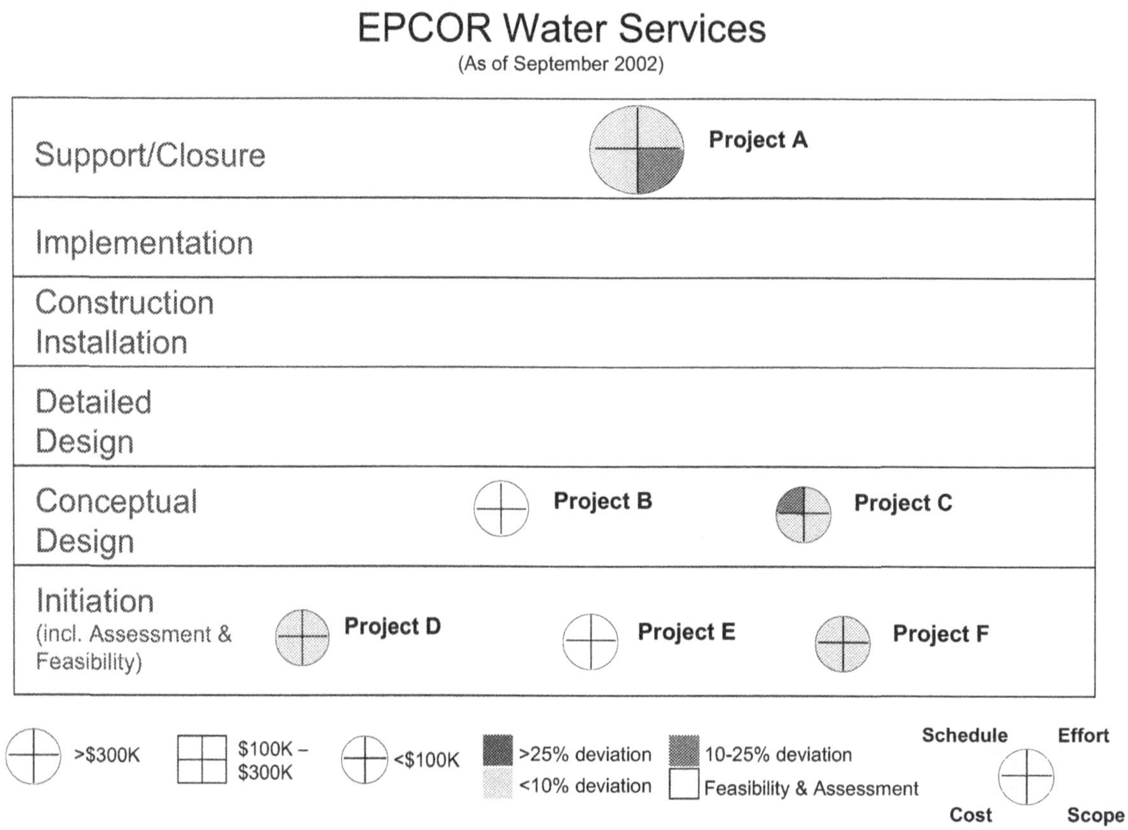


**Figure 3.37. Technology architecture roadmap guides decision**

2. Work request review – reviews a summarized list of work change requests for existing in-house tools and applications that would require less than 80 hours of IS resources. Anything over 80 hours requires the requestor to follow a more formal and detailed project request process. One benefit of reviewing work requests <80 hours is that it tends to eliminate jobs with limited value.
3. Hardware committee update – reviews planned hardware initiatives and related progress.
4. IS advisory update – reviews IT customer satisfaction statistics, analyzes customer satisfaction trends, future desktop tools and applications, and migration plans for PC upgrades.
5. Intranet online manuals – reviews progress and process.
6. IT Management council – reviews minutes, progress, and action items from the IT management council. The IT management council is chaired by the CIO and includes

senior level management from the major business units. The IT management council focuses on both IS and SCADA/DCS strategic direction and oversight.

7. Open discussion of technology – reviews and discusses new IT tools/applications and comments on current issues of existing in-house tools.

A number of subcommittees may also be actively working on various initiatives as defined by the IS steering committee. Each subcommittee is sponsored by a member of the IS steering committee. The sponsor of the subcommittee reports subcommittee progress to the IS steering committee.

***Benefits:*** The IS steering committee provides intrinsic value by bringing the major business unit managers together to drive the "spirit of corporation." Over the years, the political walls between business units have been slowing coming down. They also have been able to focus on common objectives and to leverage economy of scale between business units.

**Case Study E: Seattle Public Utilities**

| | |
|---|---|
| Services Provided | – Water Treatment/Distribution, Raw Water Transmission<br>– Wastewater Collection<br>– Stormwater Collection<br>– Solid Waste Collection/Transfer/Disposal |
| Number of Customers: | |
| Wholesale | 26 |
| Commercial | 25,533 |
| Residential | 152,589 |
| Annual Water Produced or Delivered (MG) | 44,912 |
| Annual Revenue | |
| Water | $105,345,317 |
| Total | $357,493,410 |
| Annual Utility Operating Budget | |
| Water | $53,335,768 |
| Total | $252,628,000 |
| Annual IT Operating Budget | |
| Water | $4,318,193 |
| Total | $12,130,769 |
| Utility Staffing (FTEs) | 1,281 |
| IT Staff (FTEs) | 149 |

The case study demonstrated leading practices in:

- Application Testing (Figures 3.38 and 3.39)
- Budgeting (Figures 3.40 and 3.41)
- Help Desk (Figure 3.42)
- IT Wireless (Figure 3.43 and 3.44)
- Roles and Responsibilities. (Figure 3.45 and 3.46)

These are highlighted below.

***Practice: Application Testing***

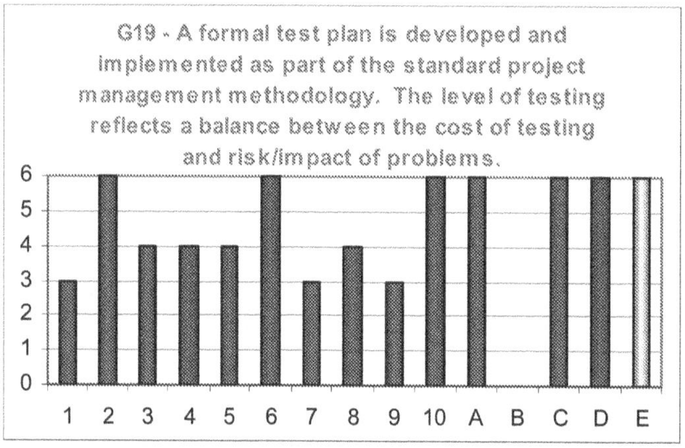


**Figure 3.38. Comparative application testing practices**

***Business Drivers:*** The IT Department had gradually received a negative perception from the utility's end users and management because of significant occurrences of problems when applications were developed and deployed. The end users encountered numerous broken programs, bugs, and other errors. The situation escalated to the point that funding for new initiatives was difficult to encumber because of the perception of project failure.

***How Achieved:*** Applications, whether written internally or purchased and implemented, are rigorously tested prior to production release. Testing follows a defined process, and is included in all projects.

Each application implementation project must include an acceptable testing plan in its initial business case analysis in order to be funded. SPU heavily promotes industry-standard testing in its culture by IT management. All IT staffers and application end-users expect and accept the need for rigorous application testing as a "normal way of doing business." Two business analysts are trained and certified in applications testing, and are assigned to each application project to design and oversee testing. Testing is phased within the project with lock-downs and freeze-dates following successful stage testing. Both "black box" and "white box" testing processes are practiced. End-users are included in white box performance and functionality testing. Applications are tested in all phases of their development and implementation. Finally, SPU includes test plans and test cases in both business unit and end-user training.

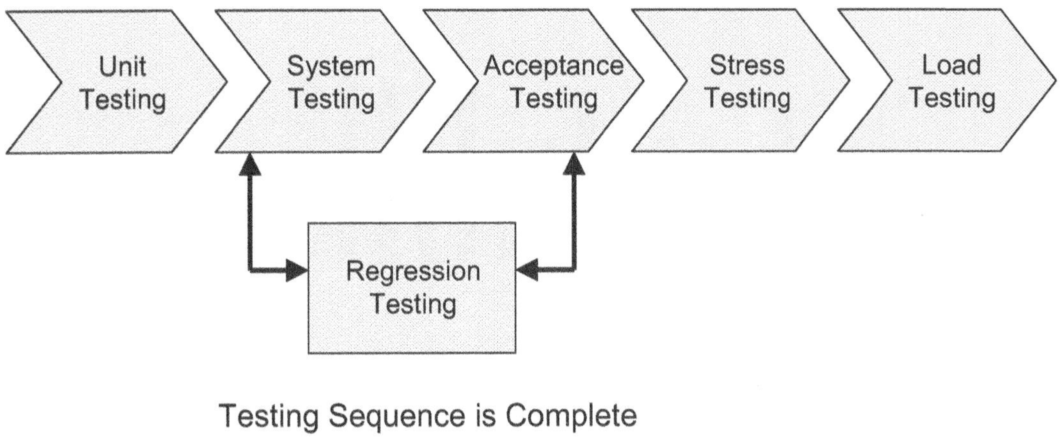


**Figure 3.39. Testing sequence**

***Benefits:*** Developing the application testing practice was time consuming and difficult to implement. The benefits, however, are now clearly recognized across the utility. One of the most significant benefits is the improved trust from management and the end-user business units in their expectations of the applications being complete, functional, and they meet the intended business need at the time of rollout.

Another benefit is the ability to get funding approval for new application development projects or application enhancement programs. In addition, the organized and systematic testing process contributes to improved development project schedule and cost management. Finally, the life cycle cost of applications is reduced when an effective testing practice is followed.

***Areas For Improvement:*** The utility believes that it should conduct a benchmarking study in order to identify and assess areas for improvement in their current application testing practice. Also, SPU has established a goal to reach one error per thousand lines of code with internally developed applications.

***Practice: Budgeting***

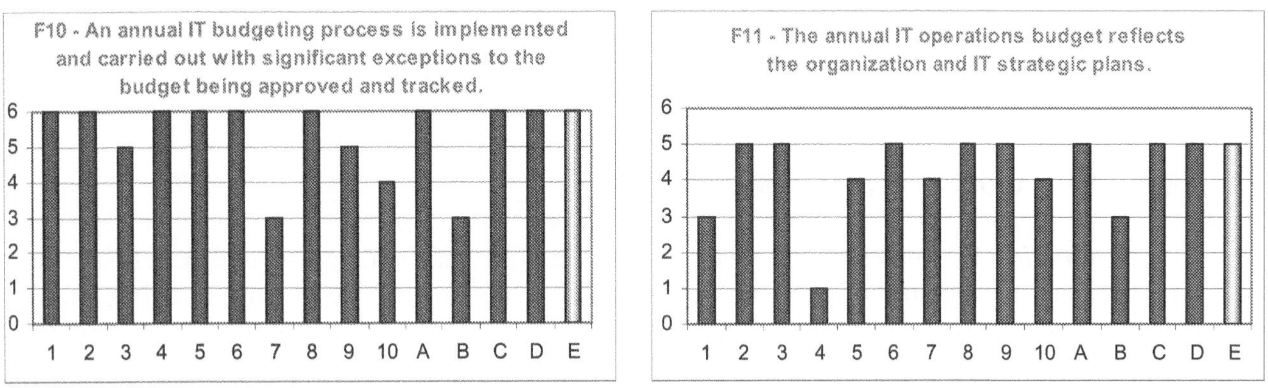


**Figure 3.40. Comparative budgeting practices**

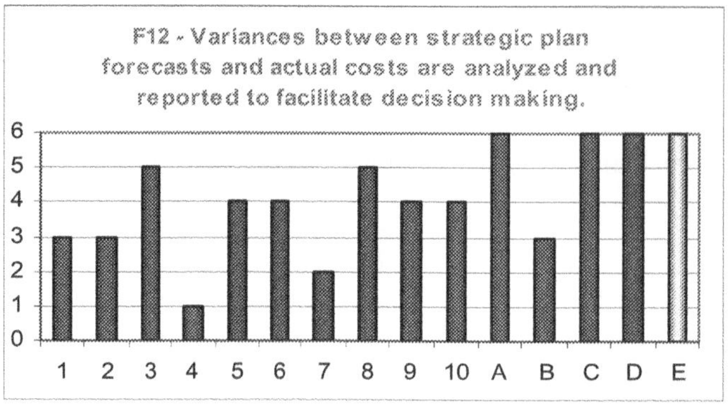


**Figure 3.40.** ***Continued***

***Business Drivers:*** The development of a well-defined budgeting process for IT application development projects is strictly followed at SPU. This practice is related to others in that it is geared toward reducing risk and improving ROI and departmental image among the utility business units.

***How Achieved:*** Project initiators (departments, user groups, etc.) must prepare and present a business case that includes a project budget, schedule, and ROI to the Technical Services Lateral Team (TSLT). The TSLT is a cross-functional governance team including business unit representatives. It stewards technology enhancements implemented by the IT Department. The TSLT must accept the budget in the plan in order for it to be approved. Medium and large projects are budgeted in "phases" to ensure accountability and budget management. Capital budgets for larger scale applications are budgeted in IT, with follow-on O&M budgets assigned to the using business unit.

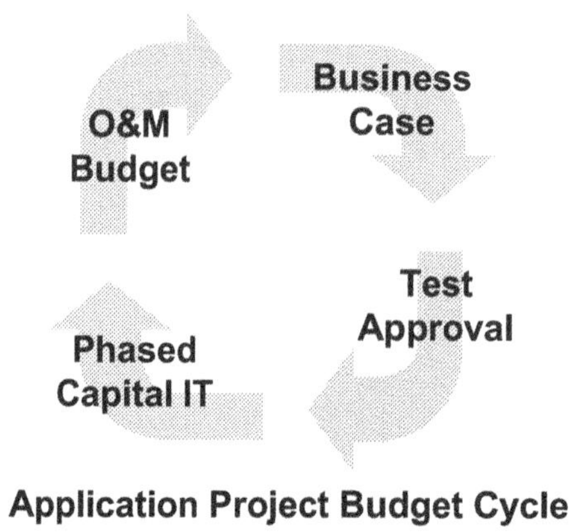


**Figure 3.41. Application project budget cycle**

***Benefits:*** By implementing this practice, the utility has realized several benefits. First, it has increased the number of application projects that are finished on budget, avoiding the cost overruns and negative perceptions resulting from previous projects. The utility is very structured in the audit process, and a benefit of this practice is that internal post project audits are easier to support and results are more favorable. And finally, the achieved Return On Investment (ROI) results in favorable "IT reputation" among the business units.

*Areas for Improvement:* The Utility believes it can improve its budgeting practice by creating a formalized process of documentation and tracking.

*Practice: Help Desk*

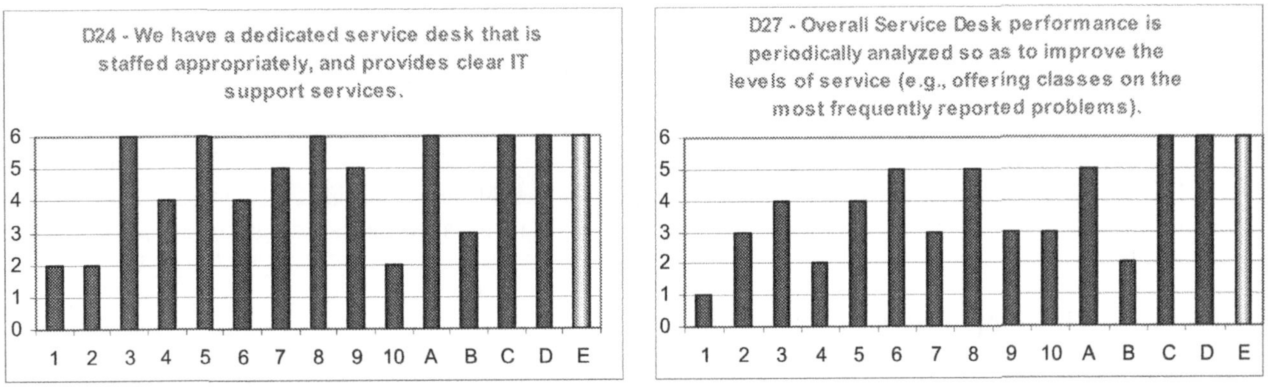


**Figure 3.42. Comparative help desk practices**

*Business Drivers:* The SPU IT Department has invested in and developed an effective Help Desk service to promote effective computing practices for its end-users. The Help Desk agents support end-users in answering questions and solving issues related to equipment, procedures, and application.

*How Achieved:* An industry-leading automated Help Desk application has been implemented for tracking, reporting, and processing calls for service (McAfee). All six full-time agents were trained on the system by the vendor's training staff. First and second level solutions are generally provided by the Help Desk agents, and higher level incidents are escalated to a dispatched technician and/or application specialist. Internal Standard Operating Procedures (SOPs) were developed and are strictly followed. Root Cause Analysis (RCA) is practiced as a tactic for preventive action. SPU has purchased and implemented a remote access tool (Windows Terminal server) to enable Help Desk agents to view and correct remote (WAN-connected) PCs. To accommodate the extended work schedules of various Utility users, the IT Department has extended the weekday service period to 12-hours (6 a.m. to 6 p.m.). In addition, 24-hour on-call support is provided using staff rotation. The Help Desk performs all PC configuration management and electronic software distribution. All PCs are imaged and "locked down" to reduce "rogue software." An annual end-user customer satisfaction survey is performed to receive and analyze feedback as a way to "stay connected" with the users' needs.

*Benefits:* By continuously striving to improve its Help Desk practice, SPU has benefited their IT service delivery in many ways. It has resulted in a significant improvement (reduction) in call volumes, reduced the number of abandoned calls, lowered user call-in wait times, and increased first call incident resolution to better than 75 percent.

*Areas For Improvement:* To continue to improve their Help Desk service, SPU is committed to providing ongoing training and certification for Help Desk staff. They will implement a feature to enable end-users to check their service ticket status online, thereby further reducing the number of calls. In addition, SPU is planning to implement an online knowledgebase of Frequently Asked Questions (FAQ) to reduce calls.

*Practice: Wireless*

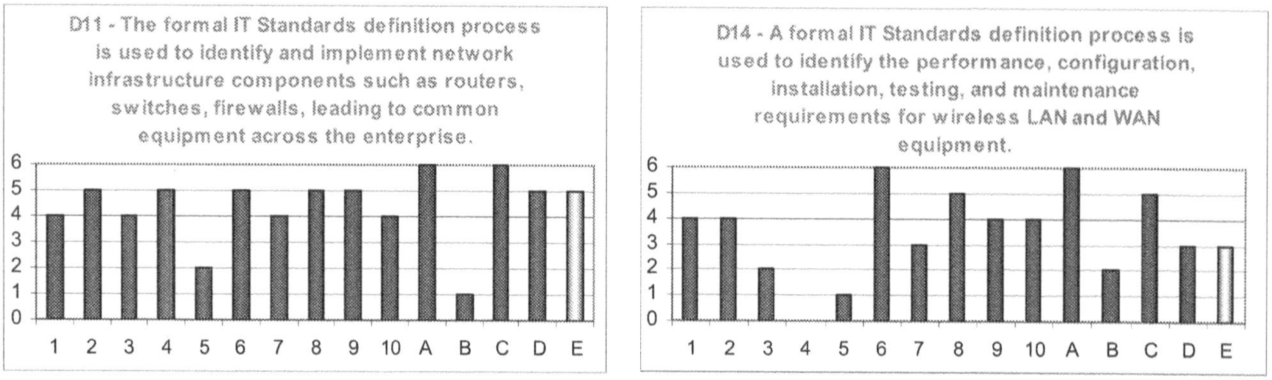


**Figure 3. 43. Comparative wireless practices**

***Business Drivers:*** SPU is piloting a wireless mobile computing service to provide selected IT services with the intention of making more information available to them and improving their workforce efficiency. Desktop applications such as e-mail, time keeping, as well as an enterprise-specific application (Computerized Maintenance Management System) are accessed via laptop computers in a mobile (vehicular) manner. Approximately 40 mobile users are being serviced in the pilot.

***How Achieved:*** Commercial Cellular Digital Packet Data (CDPD) transmission services from a national carrier are used as the wireless (radio) transport medium. Useful throughput over CDPS media is approximately 12 Kbps, and the thick-client applications were quite slow as initially rolled out. To improve performance, the utility has implemented a Citrix MetaFrame XP environment to serve the applications. A Citrix Nfuse add-on serves the content to the wireless users over a standard Internet browser.

Ruggedized laptop computers were purchased and vehicles were equipped with "quick-release" mounts to enable field users to easily remove the laptop for ease of use in the field, and wired network attachments for synchronization functions.

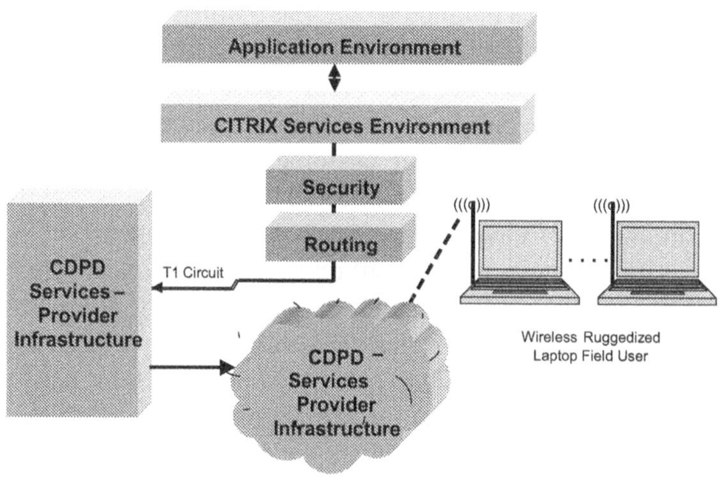


**Figure 3. 44. Example of mobile computing architecture**

***Benefits:*** SPU's mobile users are already experiencing the benefits of the technology. With the CMMS application, work orders can be accessed, completed, and updated in real time, reducing the overall time-to-complete, and increasing field worker efficiency (reducing "windshield time"). In addition, mobile users can complete their time card without having to return from the field to access a network-attached PC. Perhaps the largest benefit to date is the ability for the field workers to use e-mail for more real-time communications with others both in the field and in the wired facilities.

***Areas For Improvement:*** SPU has identified a couple of elements to implement during the Pilot rollout that will improve the service. The first is to adapt the practice and workflow associated with the CMMS application so that more benefits can be achieved from mobile computing. The second is to develop standard procedures for laptop synchronization with enterprise applications.

### *Practice: Roles and Responsibilities*

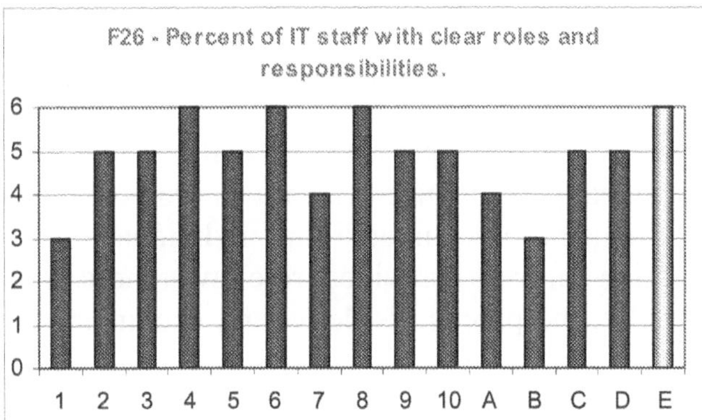


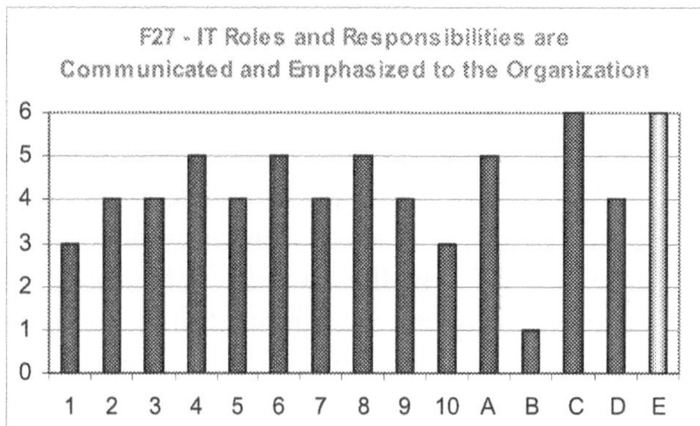


**Figure 3.45. Comparative roles and responsibilities practices**

***Business Drivers:*** Since the Utility was incorporated and formed only recently (1997) by combining separate organizations, the area of defining and establishing roles and responsibility (including IT) was critical. This area continues to be reviewed and revised regularly.

***How Achieved:*** IT uses a "service delivery and accountability" model for the organization. Since being formed in 1997, IT went through its first reorganization in 2000, with a second one in 2002. Because the utility (and IT) is a very unionized workforce, it was important to create accurate and precise definitions for titles, roles, and responsibilities. As a result of the most recent reorganization, the current IT governance structure is well defined, and is structured as shown below in Figure 3.46.

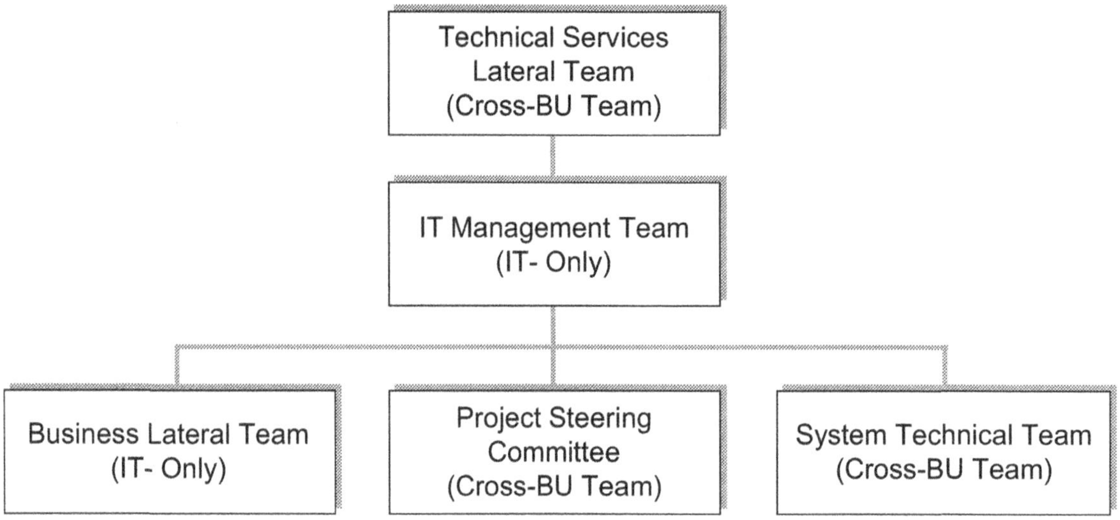


**Figure 3.46. Roles and responsibilities are well defined in IT governance structure**

***Benefits:*** SPU's IT roles and responsibilities practice has provided several benefits. For example, the concise performance agreements for annual performance evaluations promote clear expectations and measurements of roles and responsibilities. In addition, individual "self-assessments" in the annual performance evaluation process provides role awareness and accountability on an organizational level. And finally, the practice is delivering exemplary IT services in a highly unionized environment.

***Areas For Improvement:*** This governance structure, roles and responsibility definition process, and annual performance review process have been so successful, IT is driving an initiative to create the same practice in various end-user departments.

# CHAPTER 4
# APPLYING LEADING PRACTICES

## PROCESS FOR IMPROVEMENT

The metrics and IT practices survey created as part of this project provide an answer to the question "How well am I managing my information technology?" for a particular point in time. The survey results provide the foundation from which to build a plan for improving your information technology business. This chapter provides guidance about how to take the survey results and use them to create an effective improvement plan by describing the following concepts and tools:

- Foundation versus advanced information technology practices
- Information technology Practices Maturity Grid
- Action planning worksheet
- Business case worksheet
- Yearly practices survey

By applying these concepts and tools to your utility's environment, you will establish a roadmap with which to systematically and effectively upgrade your information technology practices.

## FOUNDATION AND ADVANCED INFORMATION TECHNOLOGY PRACTICES

The practice elements for each focus area can be separated into two major categories: elements that are required as a basic foundation with which to deliver information technology (foundation elements), and elements that provide advanced capabilities. Elements categorized as advanced require well-developed information technology capabilities. *If you don't have well developed foundation elements in a focus area, then the probability of success for technology projects in the advanced category is poor.*

Advanced capabilities can only be built upon a solid foundation. If for some reason it is critical to carry out practices in the advanced area while you are still working on the foundation areas (e.g., redesigning business processes as a part of new software implementation), then you should seek outside assistance to shore up your capabilities and enhance your probability of success.

Figure 4.1 shows the elements for each focus area, separated into foundation and advanced categories. Those elements that were addressed in the survey instrument are identified, as well as those elements that were addressed as part of the case studies.

In addition to the distinction between advanced and foundation elements, there are differing levels of capability within an element. These levels of capability are reflected in the survey instrument by varying scores – a score of 1 indicates a low level of capability, and a score of 6 indicates high capability. The goal is to obtain high capability in each element – that is, to obtain an average score of 6 in each focus area of the survey. *Work to get all of the foundation elements to an average score of 4.0 or higher before embarking on projects to improve advanced elements.*

**Legend for Practice Elements**

| | Element Not Surveyed | Element Surveyed | Element Surveyed with Case Study |
|---|---|---|---|
| ADVANCED | | | |
| FOUNDATION | | | |

| A. Overall Common | B. Applications and Integration | C. Data | D. Infrastructure Management | E. Service Delivery and Sourcing | F. IT Organization Direction, Oversight and Planning | G. Program and Project Management | H. Security |
|---|---|---|---|---|---|---|---|
| | | **Data Integration** | | | | | |
| **Performance Management** | | **Data Definition, Models, and Standards** | | | | | |
| **IT Staff and Development Training** | | Data Business Rules | **IT Asset Tracking (and Life-Cycle Management)** | | **Governance and Policy Definition** | | |
| Marketing | | Data Stewardship (Business Unit Role) | **Wireless (Platforms and Integration)** | | Business Process Redesign | | |
| **Customer Satisfaction** | **Application Development and Enhancement** | Data Transformation and Conversion | Telephony (IVR, ACD, Unified Messaging) | | Total Cost of Ownership | | |
| Hiring | e-Business | Data Acquisition/Input | **Network Management** | | **Strategic/ Organizational Plans** | **Testing Strategies** | |
| Organization Structure and Reporting | Integration (includes Tools and Procedures) | Data Storage | **IT Service Desk (Help Desk, Desktop Mgmt and Upgrades)** | **Service Delivery Alternatives** | **Budgeting (Capital, Operations and Annual)** | **Alignment with Strategic and Tactical Plan** | |
| Organizational Communications | Application Upgrading/ Decommissioning | Data Analysis and Reporting | Networks (Cabling, Routers, Firewalls) | Systems Development Life Cycle and Mgmt | **Roles and Responsibilities** | **Project Planning, Management and Tracking** | |
| IT Organizational Performance and Continuous Improvement | Software Applications (Selection and Implementation | Data Management | IT Architecture | **Service Level Management** | IT Vision, Mission, Goals | Program Definition and Management | Tools |
| Technical Research | Application Specific Training and Support | Data Quality | IT Security (Data, Systems, Backups) | Procurement Policies and Procedures | Business Continuity/Disaster Recovery | Quality Control/ Quality Assurance | Policies and Procedures |

**Figure 4.1. Multiple elements comprise each focus area**

An example of an advanced element would be "e-business" (electronic business). To truly offer e-business requires highly integrated applications that are available 24 hours a day, 7 days a week, and that have intensive business logic built into the applications – no human intervention required. This is clearly an advanced capability and can only be achieved with very solid infrastructure, well-established business rules, clear alignment across business systems, and very robust data transformation/conversion. By contrast, data quality is a foundation element. Without a high level of data quality (accuracy, timeliness, relevance and completeness), applications that use this data provide little or no value. Users don't trust the data, and spend time reconciling data errors. Typically, they end up getting frustrated and creating their own "shadow" systems, that they "know" are right, thus duplicating work and causing additional data reconciliation issues. Clearly, an effective information system cannot be provided without quality data.

## INFORMATION TECHNOLOGY PRACTICES MATURITY GRID

The concept of maturity levels for an organization's software process has been well developed over the past number of years. For example, the Capability Maturity Model® for Software (CMM or SW-CMM®) (Reference: SEI Pittsburgh, PA) has been developed by the software community, with stewardship by the Software Engineering Institute (SEI). This concept of process maturity involves rating software processes in terms of their levels of stability and results. The ratings start at level 0 (a non-existent process – completely reactive, with a lack of recognition of the issues this causes). Next is level 1 (ad-hoc), level 2 (repeatable but intuitive), level 3 (a defined process), level 4 (a managed and measured process), and finally, level 5 (optimized process).

The "Information Technology Practices Maturity Grid" developed as part of this project builds upon that concept and applies it not only to software practices, but all practices as assessed in the survey. The grid is presented below in Figure 4.2.

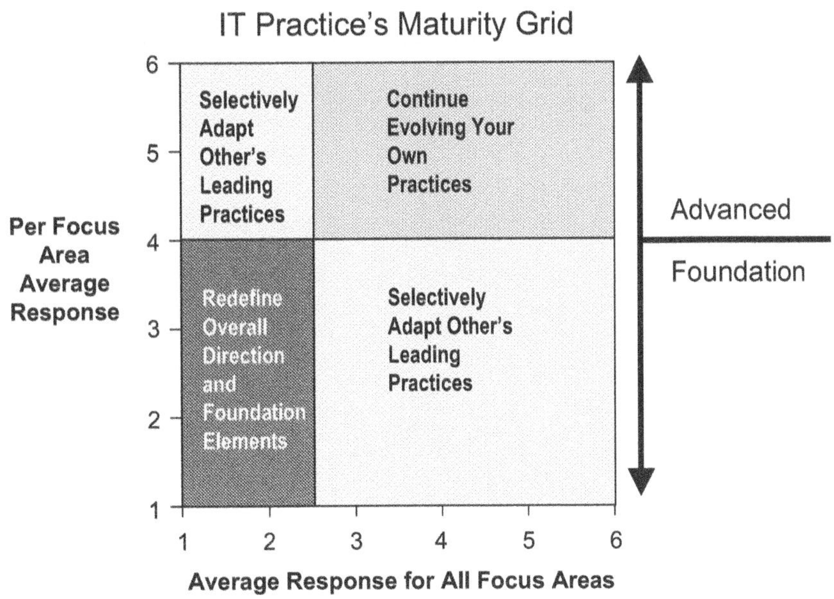


**Figure 4.2. Information Technology Practices Maturity Grid**

The grid provides guidance to utilities, based on their average response for all focus areas and their specific responses to individual focus areas. The grid is comprised of four areas, with each individual focus area plotted along a line on the grid as described below.

The first step is to map your average response for *all* focus areas on the grid, in a vertical line. Figure 4.3 shows this line plotted with an average response score of 3.5 across all focus areas.

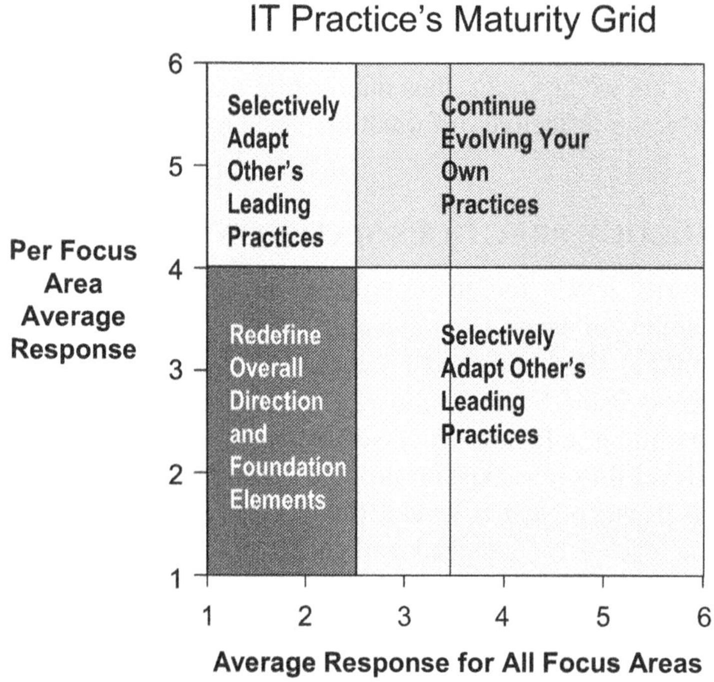


**Figure 4.3. Information technology practices maturity grid with sample average response mapped**

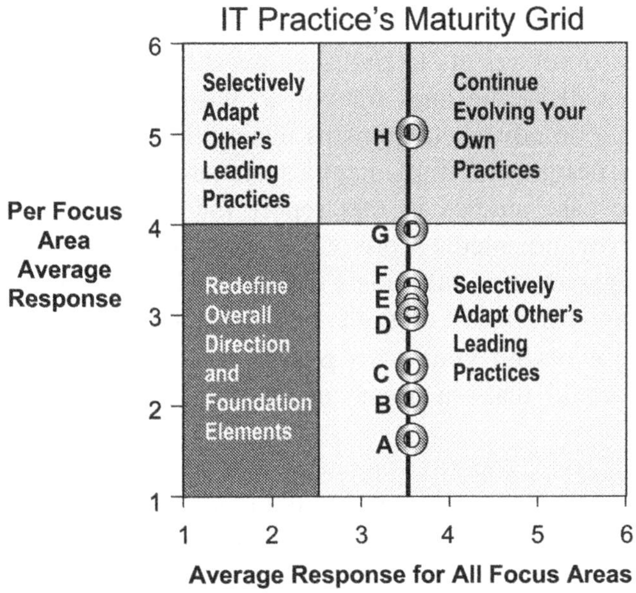


**Figure 4.4. Plotting scores on the maturity grid**

Next, take the average score for each of the eight focus areas, and plot those eight points on the vertical line (Figure 4.4). The location of each point indicates what approach to use for improving their IT practices. In this example, focus area A (Overall/Common) received an average score of 1.5 – the lowest of all the focus areas. It fell in the "Selectively adapt other's leading practices" quadrant. Focus area H (Security) received an average score of 5 – the highest score, and fell in the "Continue evolving your own practices" quadrant.

Find the lowest four to five focus area scores (in this case, focus areas A, B, C, D, and E). Identify the sector into which those focus areas fall. Use the Action Plan Worksheet (Figure 4.5) to record the lowest focus areas, and the average score of those focus areas.

The grid into which the focus areas fall determines the recommended actions, as follows:

- Grid sector "**Redefine overall direction and foundation elements.**" If you have a focus area in this sector, you should look at the foundation elements comprising that focus area. Go back to the survey, and refer to the leading practices and questions contained in that element. Use the Action Planning Worksheet to record the lowest focus areas and to select specific practices upon which to work. Once selected, systematically plan, design and implement a program to adopt the leading practice. Do not continue "business as usual" as it is typically conducted at your utility, because that will not lead you to results. You will need to adopt new ways of conducting that business practice. Refer to the case studies presented in Chapter 3 and the appendix for additional ideas. Select up to three elements upon which to work, based on your specific utilities issues and context. Implement selected practices in those elements very well, and then move on to the next set of elements within that focus area. Work on all foundation elements in all focus areas before moving to the advanced elements, unless your specific business situation indicates an urgent requirement to work on advanced elements.

- Grid sector "**Selectively adapt other's leading practices.**" If you have a focus area in this sector, use your survey results to find the lowest two or three elements within that focus area. These will be the ones for you to work on – selecting foundation elements before working on advanced elements. Using the practices identified in the survey as a guide, plan, design, and implement a systematic program to upgrade those practices. Refer to the case studies in this report for additional ideas. For those elements where the survey indicated a score of 4.0 or higher, continue doing whatever you are doing in that area. Again, only focus on two to three elements at one time – implement them well, and then move on to the next set of elements.
- Grid sector "**Continue evolving your own practices.**" Congratulations! You are doing very well and would make a great case study! Continue to develop your program.

Always work on foundation elements before working on advanced elements. If you must address some practices identified as advanced, due to business reasons or some other driving force, then be sure to acquire outside assistance, to ensure the requisite skills are available to achieve a successful outcome.

It is important to note that these guidelines must be fitted to your specific business context, and reflect your utility's culture, skills, business drivers and business understanding. For example, some factors may propel you into adopting a certain practice before you have laid the foundation to be successful in that practice. In that case, it is important to improve the odds of success by communicating the risk to other management, as well as supplementing your skills in those elements. In other cases, an IT manager or project manager with particularly solid project management skills and business acumen may be able to successfully develop an advanced element before all foundation elements are in place.

## ACTION PLANNING WORKSHEET

The worksheet shown in Figure 4.5 will assist you in selecting the specific practices upon which to focus your improvement plan. Fill out the worksheet using the guidelines stated above. Once completed, you have your list of candidate practices to consider upgrading. Examine the list, and select no more than two of the focus areas for your improvement plan. Make this selection based on your available resources, internal sponsorship, synergy with other ongoing projects, and other utility-specific factors. You should involve other people within the IT organization in making this decision, particularly those who might end up being responsible for the improvement effort.

# ACTION PLAN WORKSHEET

**Utility Name:**

| Focus Areas for Improvement (Lowest to Highest Scores) | | | |
|---|---|---|---|
| **Priority** | **Average Score** | **Focus Area** | **Leading Practices to Adapt/Adopt** |
| 1 | | | |
| | | | |
| | | | |
| 2 | | | |
| | | | |
| | | | |

**Figure 4.5. Action plan worksheet**

An example of a completed worksheet is shown in Figure 4.6. In the example, five focus areas were identified as candidates for improvement, each falling within the "selectively adapt other's leading practices." Within those focus areas, a total of twelve practices were identified as candidates. Ultimately, this fictional utility selected three of those practices to work on initially, based on an upcoming project to develop their Strategic Information Technology Plan. The three selected practices were Alignment with Strategic and Tactical Plan; IT Vision, Mission and Goals; and IT Architecture. This upcoming project provided a very good opportunity with which to develop and implement the leading practices.

# ACTION PLAN WORKSHEET

**Utility Name:** Utility USA

| **Focus Areas for Improvement (Lowest to Highest Scores)** | | | |
|---|---|---|---|
| Priority | Average Score | Focus Area | Leading Practices to Adapt/Adopt |
| **1** | **3.5** | **G** | Alignment With Strategic and Tactical Plan |
| | | **F** | IT Vision, Mission and Goals |
| | | **D** | IT Architecture |
| **2** | **3.5** | **F** | Strategic/Organizational Plans |
| | | | Governance and Policy Definition |
| | | | Total Cost of Ownership |
| **3** | **3.5** | **A** | Organizational Structure and Reporting |
| | | | IT Organizational Performance and Continuous Improvement |
| | | | |
| **4** | **3.5** | **G** | Program Definition and Management |
| | | | |
| | | | |
| **5** | **3.5** | **C** | Data Management |
| | | | Data Integration |
| | | | Data Quality |

**Figure 4.6. Example completed action plan worksheet**

## BUSINESS CASE WORKSHEET

When you have selected the two to three practices you want to improve, you will complete a "Business Case Worksheet" for each practice (Figure 4.7). This worksheet is a first step toward identifying the specific benefits that are expected from the improvement effort, the required resources, and how you will know that the practice has been improved. It addresses the following questions:

1. Why is improvement needed?
2. What will be the results/benefits? (including tangible savings)
3. How will improvement be achieved?
4. What are necessary resources and timeframe?
5. How will success be measured?

A sample worksheet is filled out for a fictional utility that has selected the Customer Service practice for improvement.

# BUSINESS CASE WORKSHEET

**Utility Name:** Utility USA

**Focus Area for Improvement:** (A) B C D E F G H
Overall/Common

**Practice to Improve:** Internal Customer Satisfaction

Why is Improvement Needed?

**Excessive number of internal customer complaints**

What Will Be the Results/Benefits? (Including tangible savings)

**Happy internal customers; increased productivity on part of our customers**

How Will Improvement Be Achieved?

**Survey customers to discover what they need/want; prioritize findings; communicate resulting actions**

What Are Necessary Resources and Timeframe?

**Two full time staff will be dedicated to complete the survey and analyze the results over a one-month period.**

How Will Success Be Measured?

**Reduction in the number of customer complaints and increased productivity among staff who deal directly with the customers.**

**Figure 4.7. Business case worksheet**

Refer to Appendix G to see example Business Case Worksheet responses by the participating utilities.

## YEARLY SURVEY AND PLANNING PROCESS

Improving IT practices requires a long-term commitment, with focus and clarity of purpose. It is easy to lose perspective over time and to either not recognize major gains, or to proclaim false victory. Thus it is key to establish a baseline and track progress on a yearly basis. The recommended steps are:

1. Conduct baseline survey.
2. Plot the results, and using the directions above, select two to three practices for an improvement during the year.
3. Integrate these practices directly into your annual IT operating plan (which should reflect your utility's annual plan).
4. Using your IT Strategic Technology Plan as a basis, develop a longer-term (three to five year) IT practices improvement plan. Ensure that the Strategic Technology Plan includes a practices component.
5. Re-do the survey on a yearly basis, comparing results by focus area and practice elements year-to-year.

Carrying out these steps will lead to a significant and measurable improvement in your ability to deliver high quality and effective information technology solutions.

## TYPICAL ISSUES AND CHALLENGES

Ensuring that value is received from your investment in information technology requires constant attention, and skilled management. A number of challenges and obstacles exist. Some of these are identified and discussed below.

- Information technology is a very immature and young field – just barely 50 years old. This has many implications, including the ongoing and continuous development of new technologies. Faster hardware, better software, wider communications pipelines, smaller and more flexible form factors, and new capabilities are developed every year. This causes misguided thinking that it is the technology that is important, rather than how technology is employed, and the changes that technology enables. People often put their focus on keeping up with hardware and software changes, rather than focusing on the real issue, which is how to get benefits from whatever hardware and software is deployed.
- When new capabilities emerge, new practices must be developed – there is no guideline for leading practices when something hasn't existed before. Hence there is a discovery period before it is clear what practices are effective. Additionally, there is no good mechanism for finding, or sharing, effective practices. It is largely an environment in which individual organizations and utilities are left to discover and create their own practices. This takes a great deal of time and effort.
- It is very easy to underestimate the cultural changes required to adopt and institutionalize new ways of working. To truly adopt new practices requires much more than just a new procedure – it requires training, communication, feedback, persuasion, leadership, vision and commitment.

## RECOMMENDATIONS

This study addresses information technology, and the associated business practices with which it is managed, selected, deployed and implemented. A number of leading practices were identified in selected areas. In addition, a survey instrument was created to allow utilities to compare their own practices to leading practices. Case studies were provided, with additional experience and perspectives on implementing selected practices. This information provides the basis for a number of recommendations for utilities desiring to improve their information technology capabilities:

1. Using the IT Practices Survey included in Appendix A, assess the practices with which you deliver information technology solutions.
2. Answer the survey questions using a cross-functional group of key users. Do not limit the perspective to people within your IT organization.
3. Plot your survey results using the tools described in this Chapter; then select two to three elements upon which to focus your improvement efforts.
4. Tie these improvement efforts to related projects in process. Develop a concrete action plan, including tasks, schedules, milestones, and expected results.
5. Track the results of the improvement efforts – if no progress is seen, analyze why and develop/implement a plan of action to address the issues.
6. When you have seen concrete positive results and they are sustained, move on to the next two to three elements upon which to focus your efforts. Continue using the survey results as a baseline and basis for making decisions.
7. Directly incorporate your survey findings and the action plan into your yearly operating plan for IT. This yearly operating plan should be a part of your Strategic IT Plan (sometimes called IT Master Plan).
8. Carry out the IT Practices Survey on a yearly basis to plot improvement trends and identify additional improvement opportunities. Plot the results to show changes on a year-to-year basis.
9. Revisit your Strategic IT Plan to ensure that it incorporates IT business practices as well as providing a technical roadmap for hardware and software investments. The Strategic IT Plan should be updated on a yearly basis; this provides a good opportunity to include the most recent findings from your yearly IT practices survey.
10. Analyze your IT budget to determine if you are investing sufficiently in your personnel and business practices.
11. Examine your efforts in the "advanced" elements – if you don't have a solid foundation upon which to build, consider whether you should delay or refocus your energy until the foundation issues have been addressed.
12. Continue to stay focused on deriving value for the IT investment, and not being distracted by the enticements of new technologies for technology's sake.

Because the science of information technology is still developing so rapidly, "leading" or "best" practices will also continue to rapidly develop. It is important to continue questing for the most effective means of ensuring that utilities receive value for the IT investments they make.

# CHAPTER 5
# ADDITIONAL NEEDS AND RESEARCH

A myriad of challenges face today's utility managers: increasing costs, new and tighter regulations, increasing customer requirements, fewer resources, and workforce development issues, to name a few. In this environment, managers cannot afford to make investments that do not successfully address at least some of these issues. Yet many managers are very skeptical about the results and benefits of their information technology investments. The results from this project show a clear way to get more from those investments. Yet this is only the beginning. There are a number of additional research needs resulting from this project.

## IDENTIFY LEADING PRACTICES FOR ALL ELEMENTS

In total, 52 elements were identified, in eight focus areas. Leading practices were developed for 40 of those elements (and are summarized in the survey), based on the participating utilities' guidance. Leading practices were not discovered for the following elements:

- Hiring
- Organization structure and reporting
- IT vision, mission, goals
- Marketing
- Business rules
- Data storage
- Data acquisition/input
- Data transformation and conversion
- Procurement policies and procedures

Yet some of these elements address issues cited by the participating utilities as being of great challenge. Developing leading practices for these elements would be of value to all utilities.

## IDENTIFY AND ANALYZE HIGH LEVERAGE LEADING PRACTICES

The practices elements were categorized into foundation elements, and advanced elements based on the research team's experience and perspective. In addition, the research team developed a hypothesis that certain foundation elements provide greater benefit than just in that element; doing them well actually enables the adoption of other leading practices and provides increased results. This hypothesis was supported by anecdotal evidence by the case studies. A valuable research project would be to determine whether there are certain practices that are essential and enabling practices. If so, these would be the most important practices upon which to focus improvement efforts, and case studies of organizations that do them well would be very valuable. A candidate list of practices for examination includes:

- Customer satisfaction
- Data analysis and reporting

- Service level management
- Strategic/organizational plans
- Alignment with strategic and tactical plan
- Project planning, management and tracking

## PROVIDE A YEARLY SURVEY AND ANALYSIS CAPABILITY

The participating utilities found great value in conducting the survey and analyzing their current practices. An idea that came up repeatedly over the course of the study involved systematizing the survey process to provide more than just a snapshot in time. This might include expanding the number of utilities taking the survey, to provide more of an industry comparison and perspective of each of the practices. Out of this larger survey, high performing utilities could be identified and used as guidance for other utilities. It would also point to industry trends regarding information technology practices, and highlight industry-wide areas for improvement.

## RESEARCH THE DISCREPANCY BETWEEN PERCEPTION OF EFFECTIVE IT SECURITY VERSUS SUPPORTING DATA

How utilities approach information technology security ("cyber-security") varies widely. In fact, the survey indicated inconsistencies between self-reported behaviors and actual data to support that behavior. At the same time that organizations reported high confidence in security practices, with very knowledgeable and well-trained staff, and strong infrastructures, they were generally unable to provide concrete data to support those perceptions. Given the high level of utility reliance on automation today, and the criticality of information technology in many utilities, this apparent discrepancy should be resolved. Specifically, how utilities are currently ensuring their security information technology practices should be researched in more depth.

## DEFINE AN EFECTIVE PROGRAM TO IMPROVE DATA PRACTICES

Based on the survey results, most utilities are struggling with effective management of data. On average, the data focus area scores were the lowest of any focus area. Inconsistent, lacking, or incorrect data jeopardizes any computer application and threatens user confidence in the system. Yet there is no well-known and clear roadmap for utilities to adopt, should they wish to embark on an improvement program in this area. Further developing guidelines, policies and activities with which to improve data management would be of value, since data is at the heart of every single IT system.

## EXAMINE RECENT SUCCESSFUL AND FAILED INFORMATION TECHNOLOGY IMPLEMENTATIONS

More than 100 effective IT practices were identified as part of this project. These practices should be further analyzed to ensure a high relationship between successful implementations, and a low relationship to unsuccessful implementations. This can be done through a review of recent utility implementations, both successful and unsuccessful. Surveying utilities with recent implementations regarding their practices would provide a base of information for analyzing the relationship of leading practices to successful implementations. This would be of value to all utilities about to upgrade or implement new technology solutions.

## VALIDATE THE HYPOTHESES REGARDING LEVELS OF INVESTMENT IN INFORMATION TECHNOLOGY

Two hypotheses were developed as a result of analyzing the utility survey data and comparing it to the case study data.

1. You do not necessarily have to spend additional money to achieve leading practices in IT. The leading practices organizations spent less than 5 percent of their operating budget on IT.
2. You may need to spend more for sufficient and skilled IT staff and reduce spending on other IT resources (such as hardware and software). The leading practice organizations IT staff averaged almost 10 percent of the total company full-time equivalents (FTEs). They also demonstrated a clear focus on changing or redesigning their internal practices using existing tools to gain efficiencies. This required that the IT staff provide more than just technical prowess, but also provide capabilities such as business understanding, effective communications, negotiation skills, meeting facilitation, and so on.

These hypotheses have important implications regarding how much utilities invest in information technology, and how they approach IT staffing and associated roles and responsibilities.

A research project that provided guidance regarding appropriate levels of spending for IT, and that provided guidance about IT staffing issues, would be of high interest and high value to all utility managers.

## UTILITY PROVIDED APPLICATION-SPECIFIC RESEARCH IDEAS

During the project, the participating utilities shared information amongst themselves regarding recent, current and future applications implementation projects. They identified the following areas as of particular interest: customer information systems (customer service, billing collections, automated meter reading, call tracking/dispatching, service orders, interactive voice response); geographic information systems; enterprise decision support systems; document management (including records and knowledge management), finance (general ledger, accounts receivable, accounts payable, budgeting, purchasing, contract management) and computerized maintenance management/work management systems.

As they shared information on these topics, they identified the following research items of interest:

- Customer Information Systems (CIS). This application is viewed as complex, high risk, with a fragmented vendor market, and significant customization needs. There is strong concern about the risk involved in selecting and implementing a new system, or even updating the current (often homegrown) system. They would like guidance regarding an effective selection and implementation process for a new CIS, or for any high-risk application. They would also like a survey of other utility experiences in moving to a new CIS, and various approaches including vendor selection, and whether or not to outsource.

- Enterprise Decision Support Systems. Utilities are struggling with lots of data available, but little information. They are also struggling with managing performance, and how enterprise systems support this effort.
- Document Management Systems. They would like guidance regarding requirements definition, document conversion, how to address records management, what the role of knowledge management should be, and whether an integrated solution for both document and content management is appropriate. Document management systems projects are viewed as requiring a big culture change – how best should utilities do this?
- Budgeting. These mostly manual systems are slow and have significant redundancy. They do not provide access to current data for managers or operating personnel. Utilities would like guidance regarding how to make the required internal business process changes, particularly when there is a historic distrust of IT combined with untrained, reluctant finance, accounting, and budgeting staff.
- All systems. They would like guidance regarding how to set realistic expectations and establish alignment between users, technology providers, and management. They are interested in finding successful implementations that can be used as models from which to build their own implementation plans. And they are interested in business case guidance – how can they, and should they, show that a new application should or should not be invested in?

# APPENDIX A
# SAMPLE SURVEY, PARTS I AND II

Following are "snapshots" of Part I and II of the survey.

## Part I – Metric and Profile

**C. UTILITY CUSTOMERS:**

| | | | Year X | Year Y | Year Z | 3 Year Avg. | A | E | R |
|---|---|---|---|---|---|---|---|---|---|
| **Potable Water Customers:** | | (Number of Accounts) | | | | | | | |
| | 1 | Wholesale #: | ____ | ____ | ____ | ____ | ○ | ○ | ○ |
| | 2 | Commercial/Industrial #: | ____ | ____ | ____ | ____ | ○ | ○ | ○ |
| | 3 | Residential #: | ____ | ____ | ____ | ____ | ○ | ○ | ○ |

A=Accurate; E=Estimated; R=Rough Guess

**D. ANNUAL WATER PRODUCED OR DELIVERED:**

| | | | Year X | Year Y | Year Z | 3 Year Avg. | A | E | R |
|---|---|---|---|---|---|---|---|---|---|
| **Water (Potable)** | 1 | MG/Year: | ____ | ____ | ____ | ____ | ○ | ○ | ○ |

A=Accurate; E=Estimated; R=Rough Guess

**E. ANNUAL UTILITY REVENUE:**

| | | | | | | | A | E | R |
|---|---|---|---|---|---|---|---|---|---|
| **Water (Potable)** | 1 | (US Dollars) | ____ | ____ | ____ | ____ | ○ | ○ | ○ |
| **All Other** | 2 | (US Dollars) | ____ | ____ | ____ | ____ | ○ | ○ | ○ |
| **TOTAL:** | 3 | (US Dollars) | ____ | ____ | ____ | ____ | ○ | ○ | ○ |

**F. ANNUAL UTILITY OPERATING BUDGET:**

Includes all utility operations and maintenance spending other than capital (i.e., excludes debt service, taxes, franchise fees, payments in lieu of taxes to municipalities, etc.). Also includes direct costs or allocated costs for all business services including IT, HR, Finance, Procurement, Lab, Fleet, Engineering, Facilities, Customer Service, Legal, Public Relations, etc.).

| | | | | | | | A | E | R |
|---|---|---|---|---|---|---|---|---|---|
| **Water (Potable)** | 1 | (US Dollars) | ____ | ____ | ____ | ____ | ○ | ○ | ○ |
| **All Other** | 2 | (US Dollars) | ____ | ____ | ____ | ____ | ○ | ○ | ○ |
| **TOTAL:** | 3 | (US Dollars) | ____ | ____ | ____ | ____ | ○ | ○ | ○ |

## Part II – Practice Questions

| B.14. A formal process is in place to train application developers in the effective use of the development tools and methodologies. | |
|---|---|
| No formal approach – 1 | ○ |
| Formality as needed; based on criticality – 2 | ○ |
| Sometimes formal procedure: an aspect of some situations – 3 | ○ |
| Typically formal; an aspect of most situations – 4 | ○ |
| Often do this, omitted only in exceptional circumstances – 5 | ○ |
| Always do this: standard operating procedure – 6 | ○ |

| B.6. Business processes are streamlined prior to new application implementation, and reviewed for effectiveness subsequent to the implementation. | |
|---|---|
| No results seen – 1 | ○ |
| Minimal results; long way to go – 2 | ○ |
| Some results; still below expectations – 3 | ○ |
| Good results; getting there – 4 | ○ |
| Excellent results; still some room to improve – 5 | ○ |
| Unparalleled results; a total success – 6 | ○ |

# APPENDIX B
# SURVEY RESPONSE SUMMARIES

## METRICS (PROFILE) RESPONSES

This appendix provides a summary of the survey responses, including an overview of the responses to metrics questions and an overview of utility responses to each of the practice focus areas.

In general, the specific survey question or field being referenced is depicted directly on the chart or summary. For example, chart 1 shows the total number of utility customers for each of the respondents. This is defined as the sum of field C1 (number of wholesale water accounts) + C2 (number of commercial/industrial accounts) + C3 (number of residential accounts). Where relevant, the unit is also provided (K equals thousands and M equals millions). All dollars have been converted to US currency.

A total of fifteen (15) organizations completed the survey. Identifiers A through E show the benchmarking partners that also include EPCOR and Seattle Public Utilities. Identifiers 1 through 11 show the participating utilities. Note that two of the participating utilities were eventually selected to be benchmarking partners; they are included in the A through E designations. One participating utility (11) was unable to commit the resources for the entire project; their data is included where available.

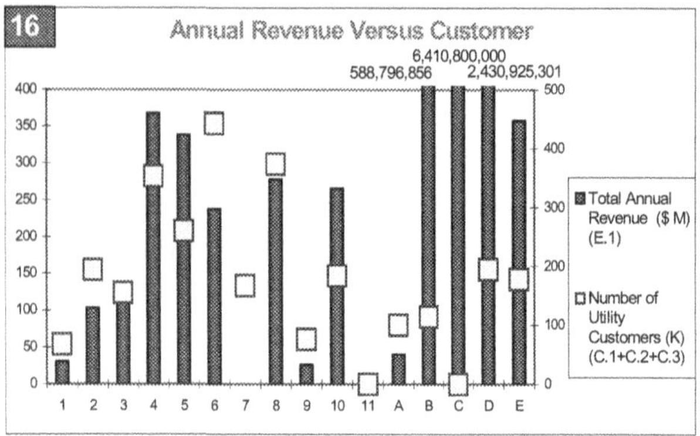
16
Annual Revenue Versus Customer
6,410,800,000
588,796,856
2,430,925,301
Total Annual Revenue ($ M) (E.1)
Number of Utility Customers (K) (C.1+C.2+C.3)

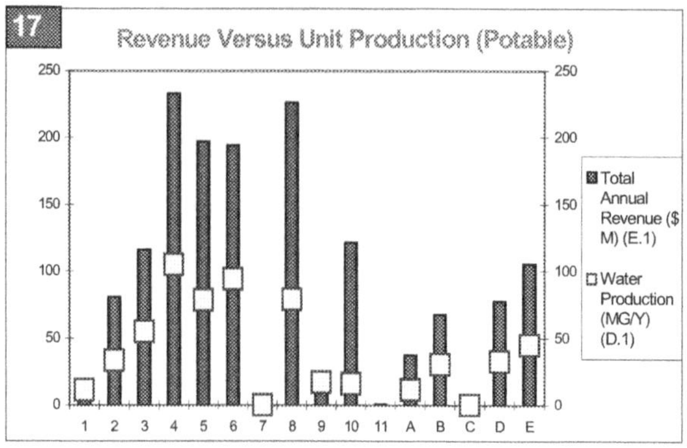
17
Revenue Versus Unit Production (Potable)
Total Annual Revenue ($ M) (E.1)
Water Production (MG/Y) (D.1)

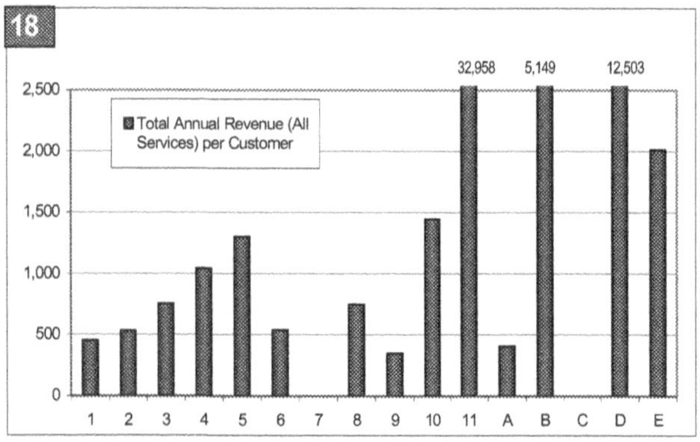
18
32,958
5,149
12,503
Total Annual Revenue (All Services) per Customer

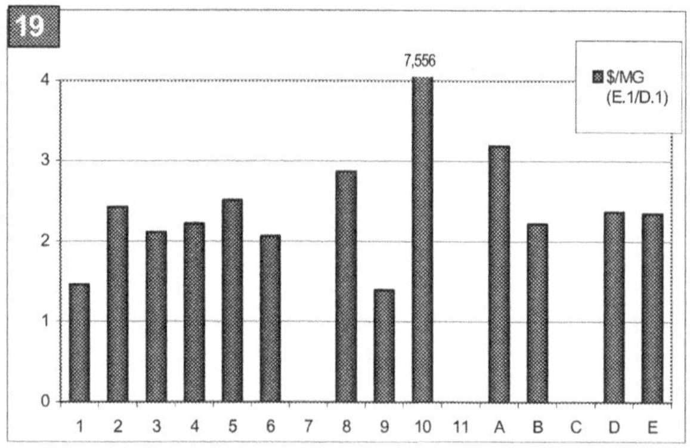
19
7,556
$/MG (E.1/D.1)

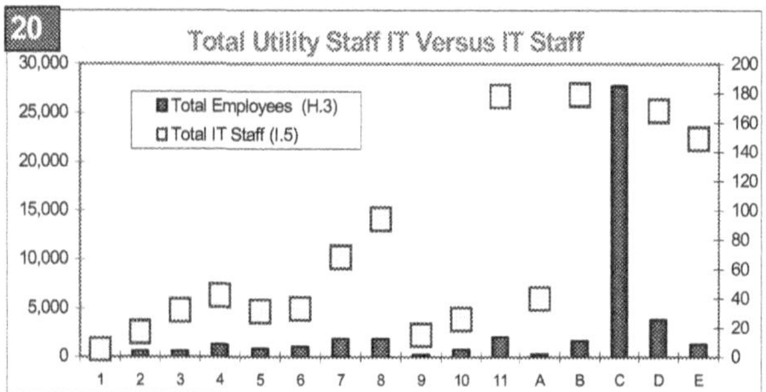
20
Total Utility Staff IT Versus IT Staff
Total Employees (H.3)
Total IT Staff (I.5)

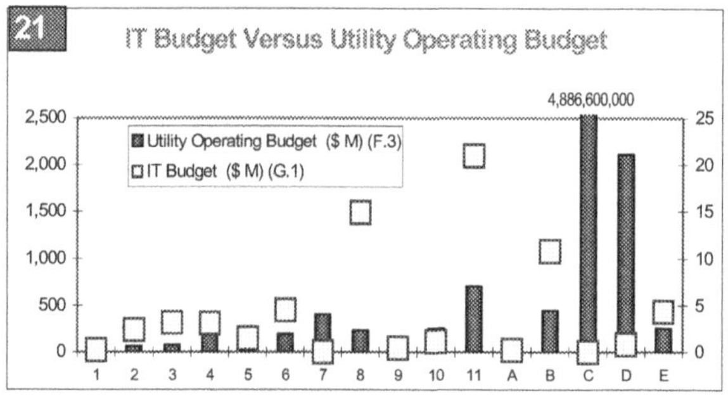
21
IT Budget Versus Utility Operating Budget
4,886,600,000
Utility Operating Budget ($ M) (F.3)
IT Budget ($ M) (G.1)

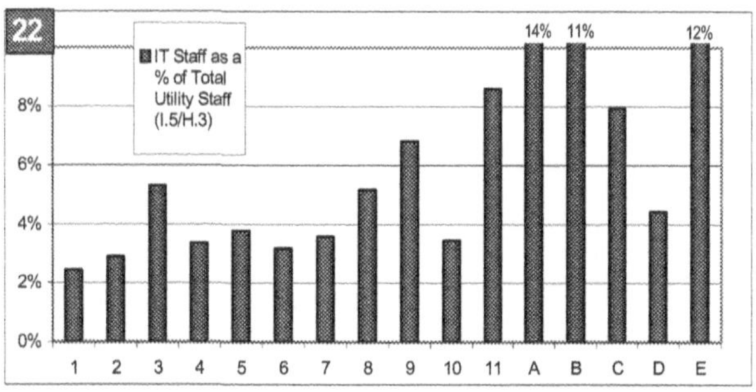
22
IT Staff as a % of Total Utility Staff (I.5/H.3)
14%
11%
12%

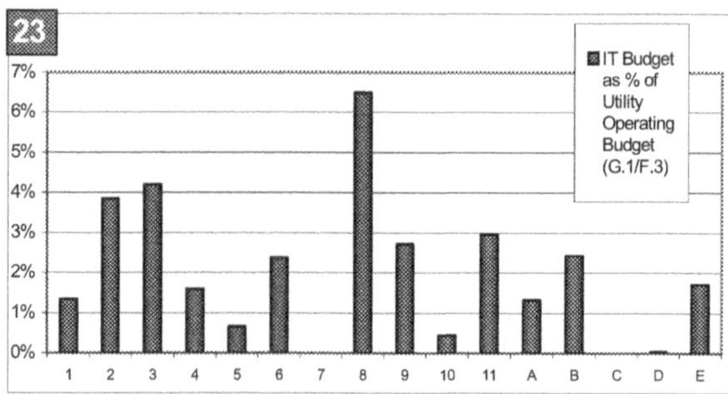
23
IT Budget as % of Utility Operating Budget (G.1/F.3)

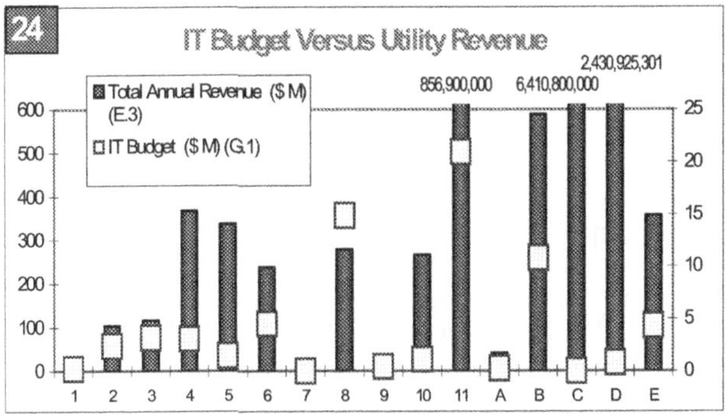
24
IT Budget Versus Utility Revenue
Total Annual Revenue ($ M) (E.3)
IT Budget ($ M) (G.1)
856,900,000
6,410,800,000
2,430,925,301

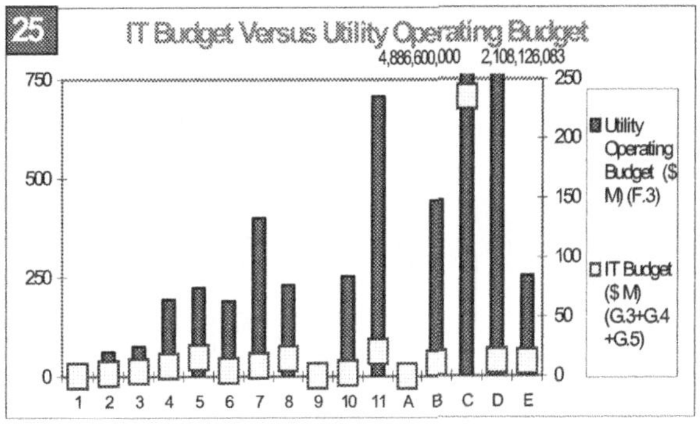
25
IT Budget Versus Utility Operating Budget
4,886,600,000
2,108,126,083
Utility Operating Budget ($ M) (F.3)
IT Budget ($ M) (G.3+G.4+G.5)

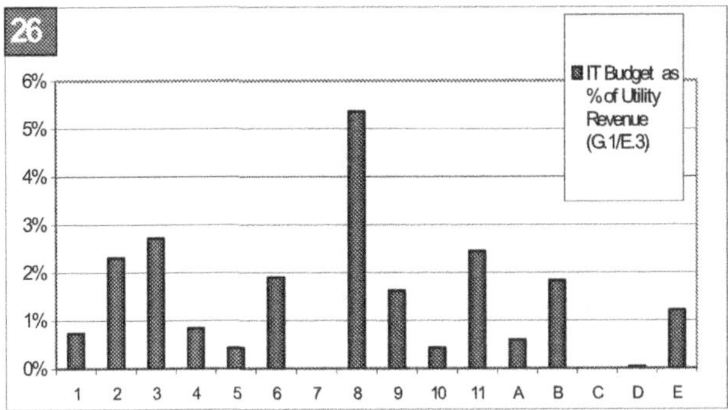
26
IT Budget as % of Utility Revenue (G.1/E.3)

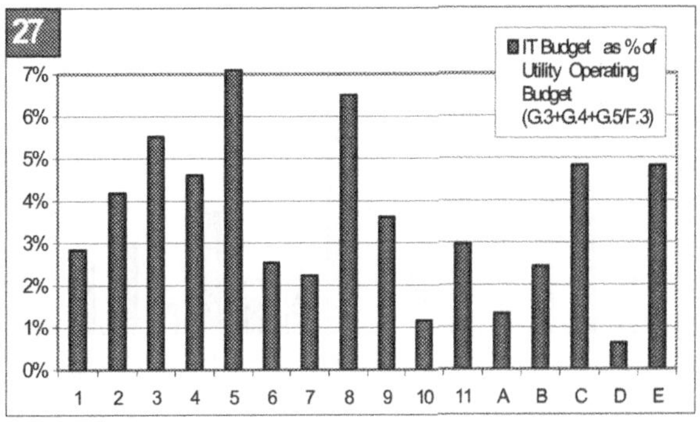
27
IT Budget as % of Utility Operating Budget (G.3+G.4+G.5/F.3)

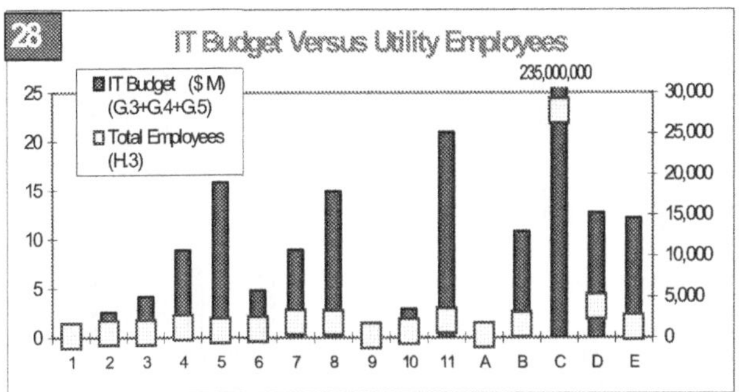
28
IT Budget Versus Utility Employees
IT Budget ($ M) (G.3+G.4+G.5)
Total Employees (H.3)
235,000,000

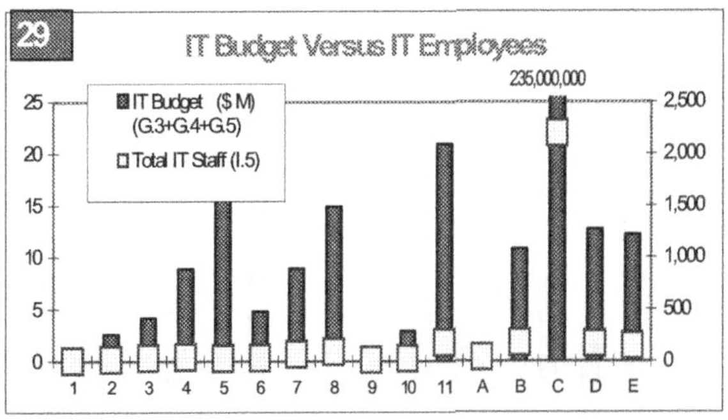
29
IT Budget Versus IT Employees
IT Budget ($ M) (G.3+G.4+G.5)
Total IT Staff (I.5)
235,000,000

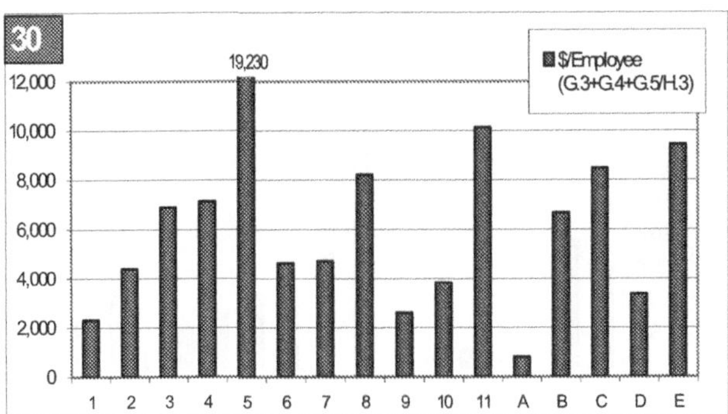
30
19,230
$/Employee (G.3+G.4+G.5/H.3)

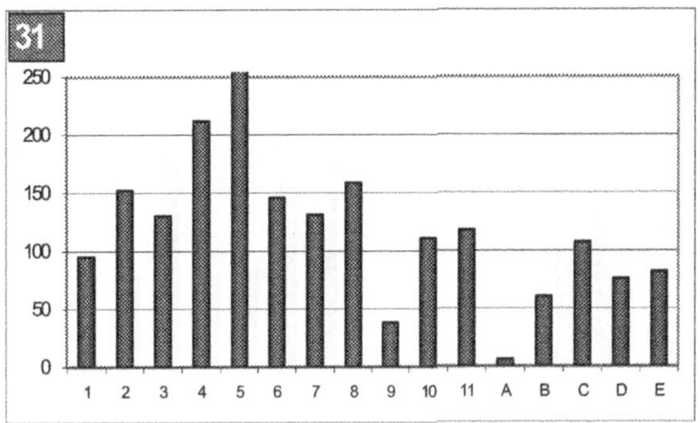
31

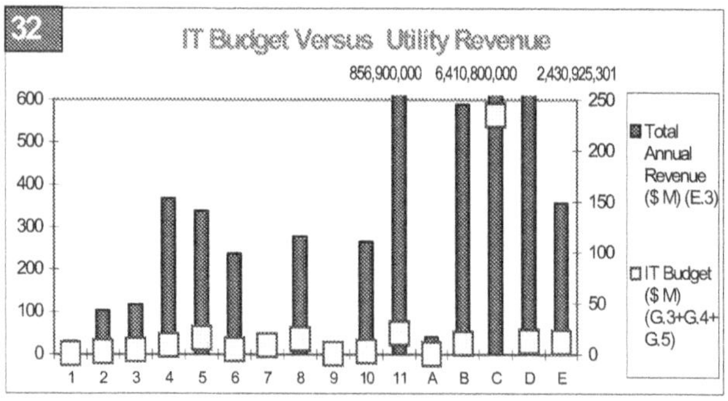
32
IT Budget Versus Utility Revenue
856,900,000 6,410,800,000 2,430,925,301
Total Annual Revenue ($ M) (E.3)
IT Budget ($ M) (G.3+G.4+G.5)
1 2 3 4 5 6 7 8 9 10 11 A B C D E

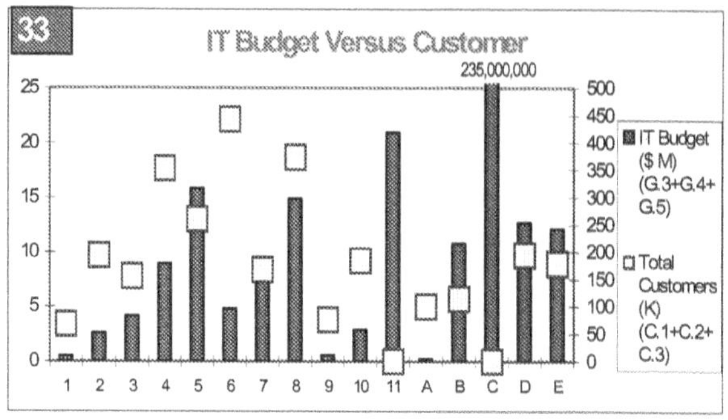
33
IT Budget Versus Customer
235,000,000
IT Budget ($ M) (G.3+G.4+G.5)
Total Customers (K) (C.1+C.2+C.3)
1 2 3 4 5 6 7 8 9 10 11 A B C D E

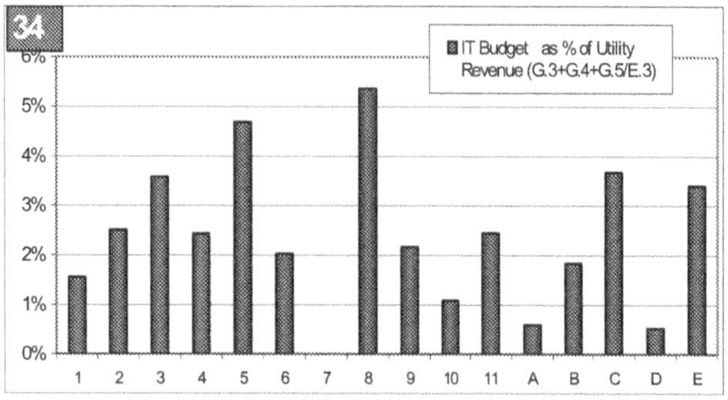
34
IT Budget as % of Utility Revenue (G.3+G.4+G.5/E.3)
1 2 3 4 5 6 7 8 9 10 11 A B C D E

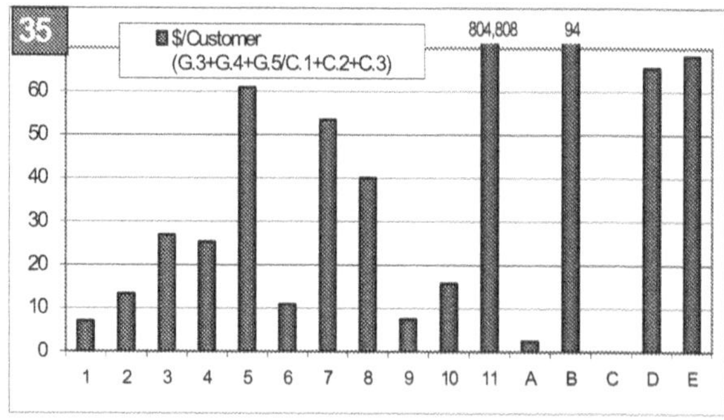
35
$/Customer (G.3+G.4+G.5/C.1+C.2+C.3)
804,808 94
1 2 3 4 5 6 7 8 9 10 11 A B C D E

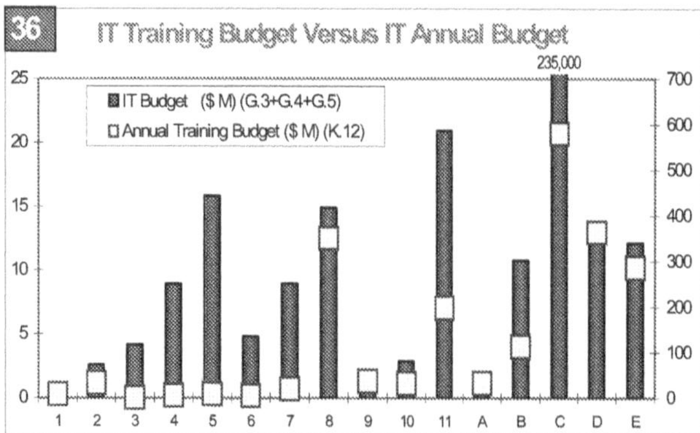
36
IT Training Budget Versus IT Annual Budget
235,000
IT Budget ($ M) (G.3+G.4+G.5)
Annual Training Budget ($ M) (K.12)
1 2 3 4 5 6 7 8 9 10 11 A B C D E

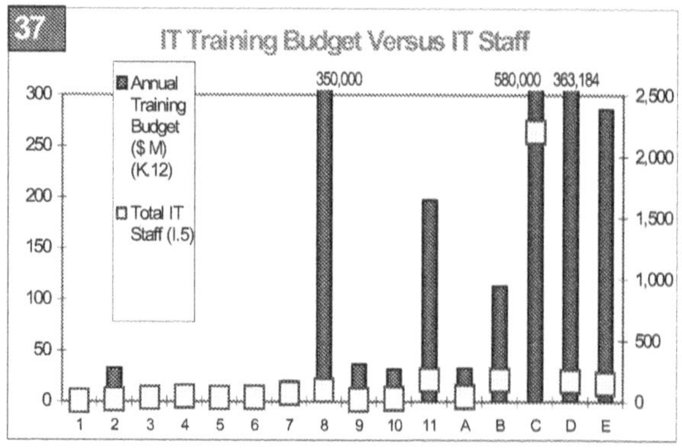
37
IT Training Budget Versus IT Staff
350,000 580,000 363,184
Annual Training Budget ($ M) (K.12)
Total IT Staff (I.5)
1 2 3 4 5 6 7 8 9 10 11 A B C D E

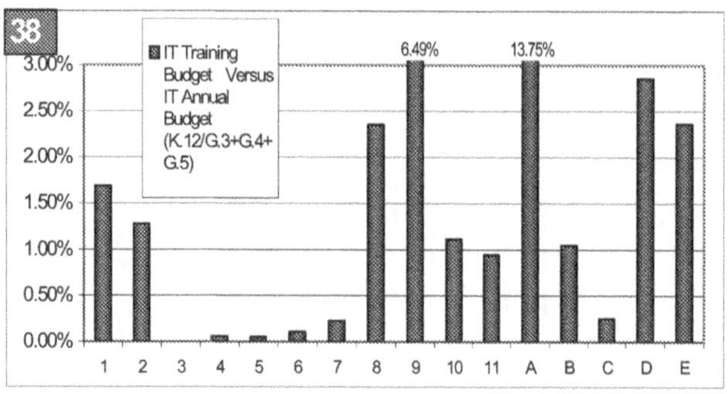
38
IT Training Budget Versus IT Annual Budget (K.12/G.3+G.4+G.5)
6.49% 13.75%
1 2 3 4 5 6 7 8 9 10 11 A B C D E

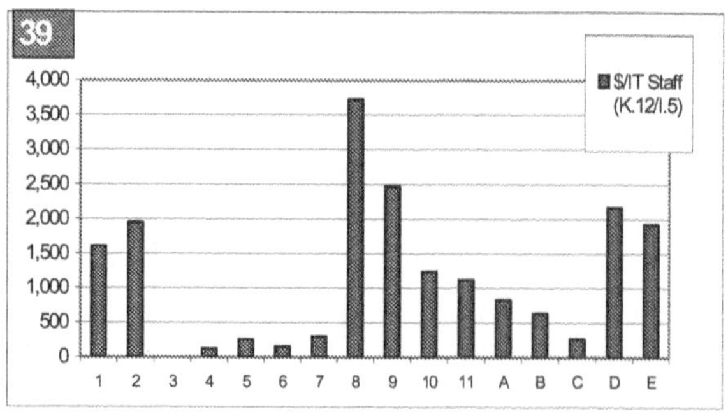
39
$/IT Staff (K.12/I.5)
1 2 3 4 5 6 7 8 9 10 11 A B C D E

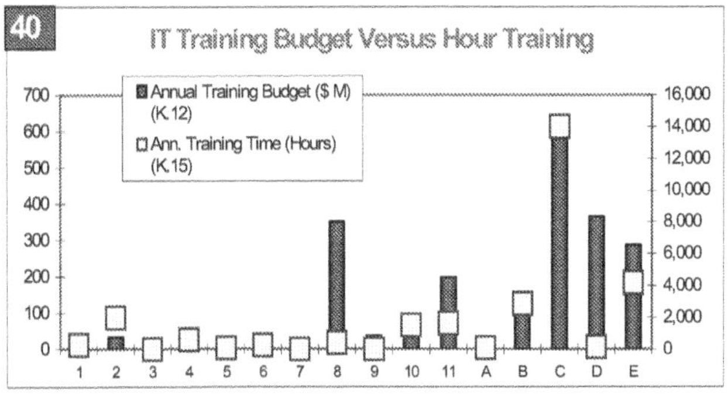
40
IT Training Budget Versus Hour Training
Annual Training Budget ($ M) (K.12)
Ann. Training Time (Hours) (K.15)
700 600 500 400 300 200 100 0
16,000 14,000 12,000 10,000 8,000 6,000 4,000 2,000 0
1 2 3 4 5 6 7 8 9 10 11 A B C D E

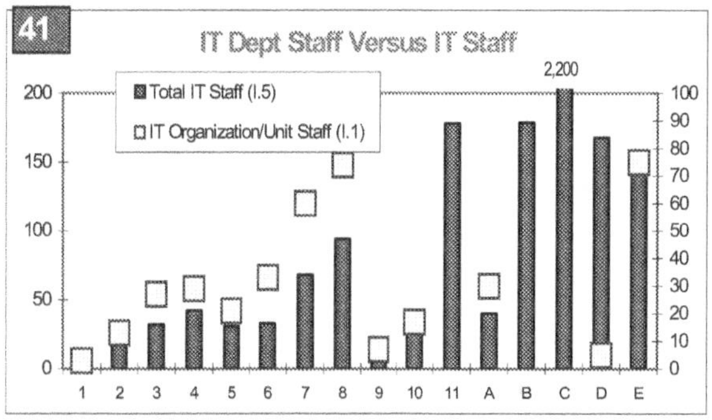
41
IT Dept Staff Versus IT Staff
Total IT Staff (I.5)
IT Organization/Unit Staff (I.1)
2,200
200 150 100 50 0
100 90 80 70 60 50 40 30 20 10 0
1 2 3 4 5 6 7 8 9 10 11 A B C D E

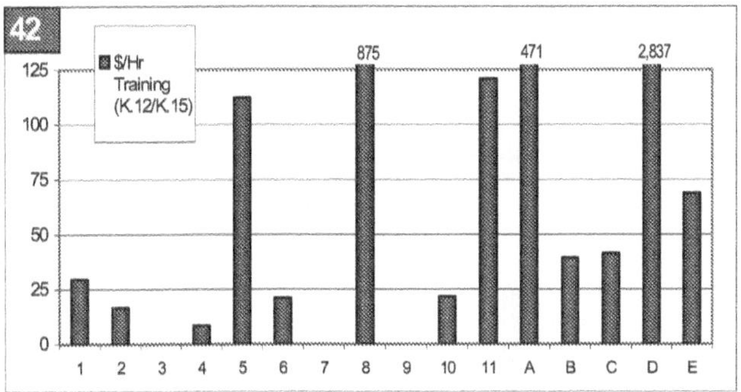
42
$/Hr Training (K.12/K.15)
875
471
2,837
125 100 75 50 25 0
1 2 3 4 5 6 7 8 9 10 11 A B C D E

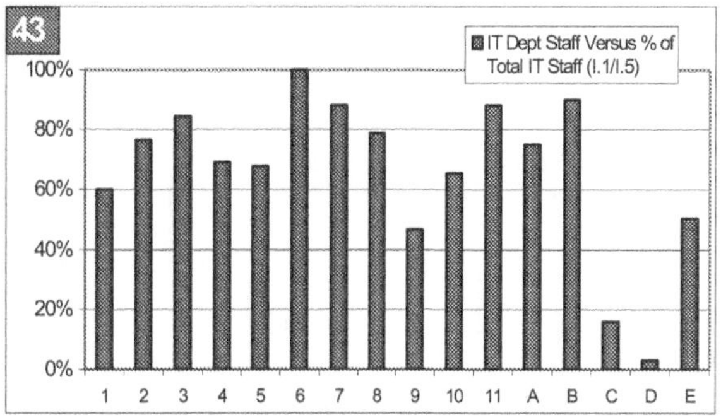
43
IT Dept Staff Versus % of Total IT Staff (I.1/I.5)
100% 80% 60% 40% 20% 0%
1 2 3 4 5 6 7 8 9 10 11 A B C D E

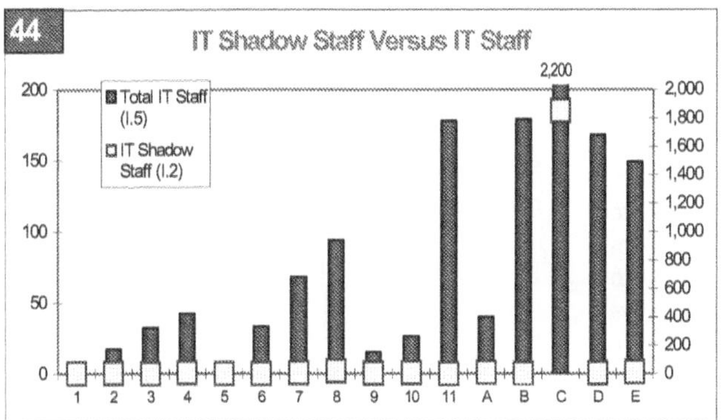
44
IT Shadow Staff Versus IT Staff
Total IT Staff (I.5)
IT Shadow Staff (I.2)
2,200
200 150 100 50 0
2,000 1,800 1,600 1,400 1,200 1,000 800 600 400 200 0
1 2 3 4 5 6 7 8 9 10 11 A B C D E

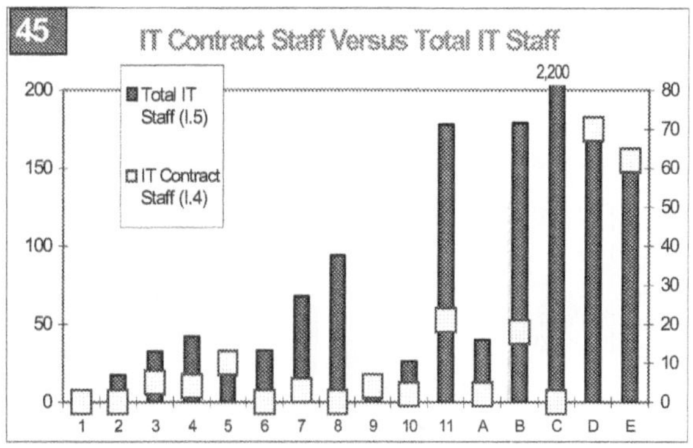
45
IT Contract Staff Versus Total IT Staff
Total IT Staff (I.5)
IT Contract Staff (I.4)
2,200
200 150 100 50 0
80 70 60 50 40 30 20 10 0
1 2 3 4 5 6 7 8 9 10 11 A B C D E

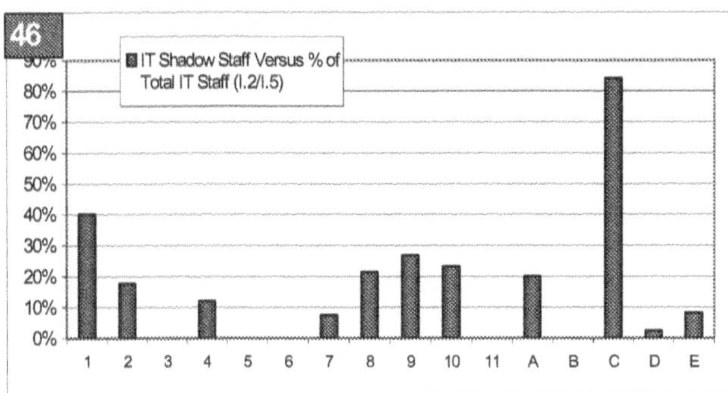
46
IT Shadow Staff Versus % of Total IT Staff (I.2/I.5)
80% 70% 60% 50% 40% 30% 20% 10% 0%
1 2 3 4 5 6 7 8 9 10 11 A B C D E

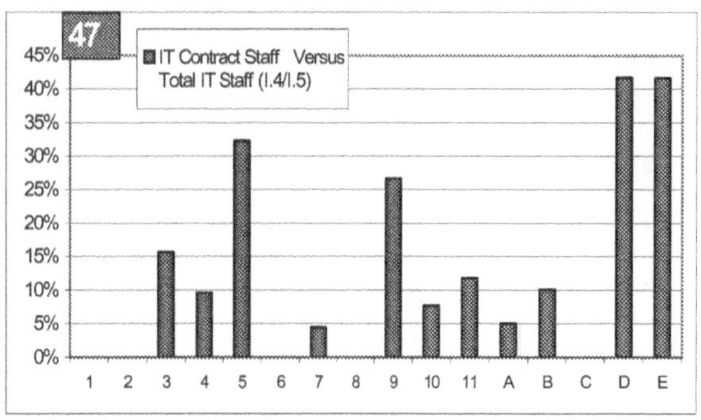
47
IT Contract Staff Versus Total IT Staff (I.4/I.5)
45% 40% 35% 30% 25% 20% 15% 10% 5% 0%
1 2 3 4 5 6 7 8 9 10 11 A B C D E

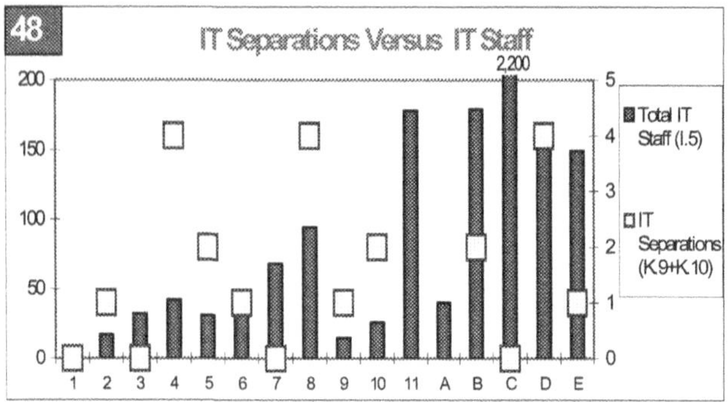
48
IT Separations Versus IT Staff
2,200
Total IT Staff (I.5)
IT Separations (K.9+K.10)

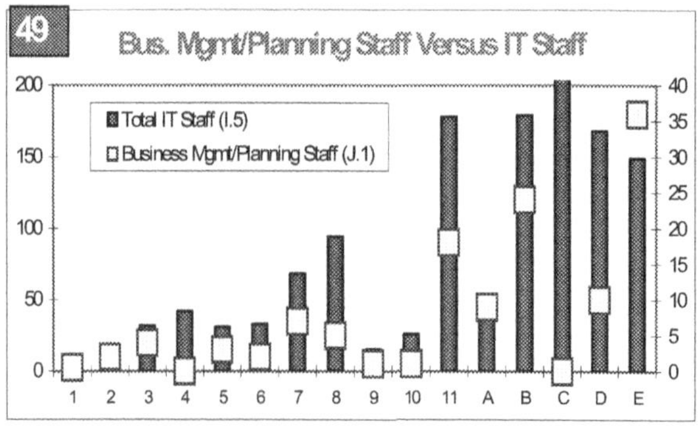
49
Bus. Mgmt/Planning Staff Versus IT Staff
Total IT Staff (I.5)
Business Mgmt/Planning Staff (J.1)

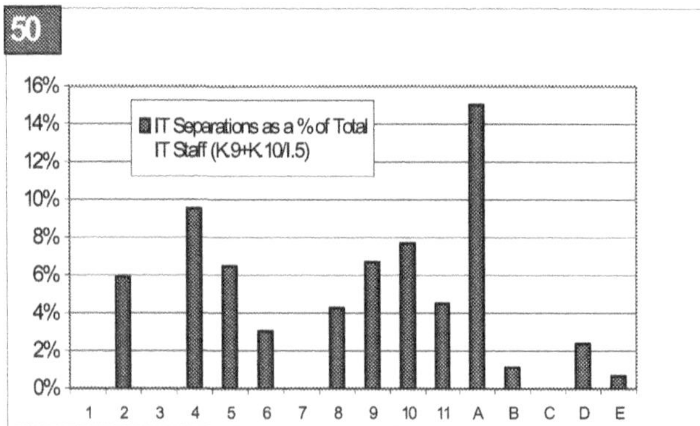
50
IT Separations as a % of Total IT Staff (K.9+K.10/I.5)

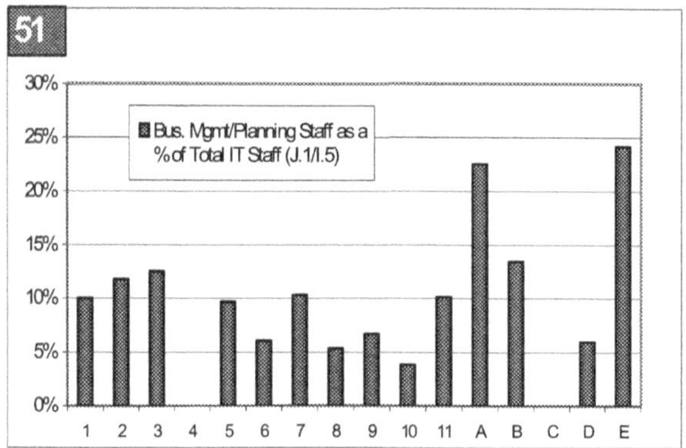
51
Bus. Mgmt/Planning Staff as a % of Total IT Staff (J.1/I.5)

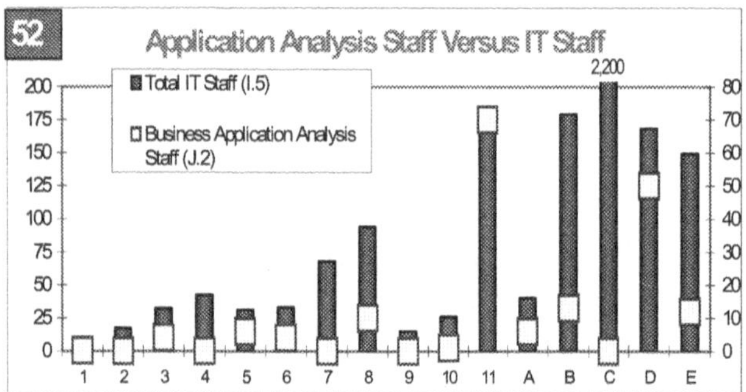
52
Application Analysis Staff Versus IT Staff
2,200
Total IT Staff (I.5)
Business Application Analysis Staff (J.2)

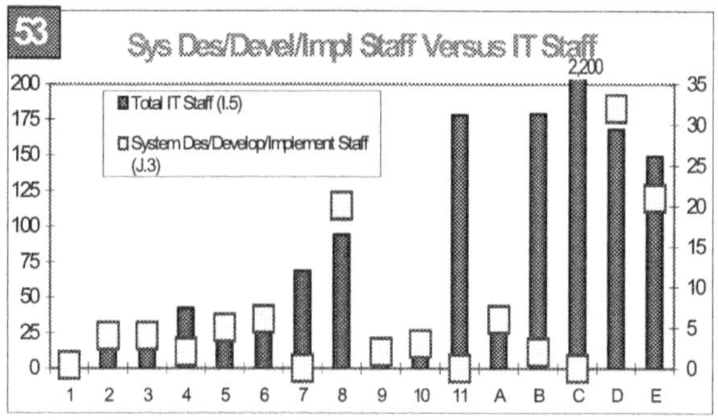
53
Sys Des/Devel/Impl Staff Versus IT Staff
2,200
Total IT Staff (I.5)
System Des/Develop/Implement Staff (J.3)

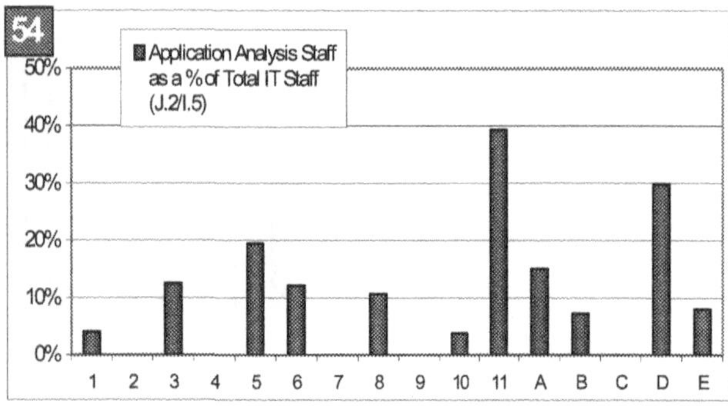
54
Application Analysis Staff as a % of Total IT Staff (J.2/I.5)

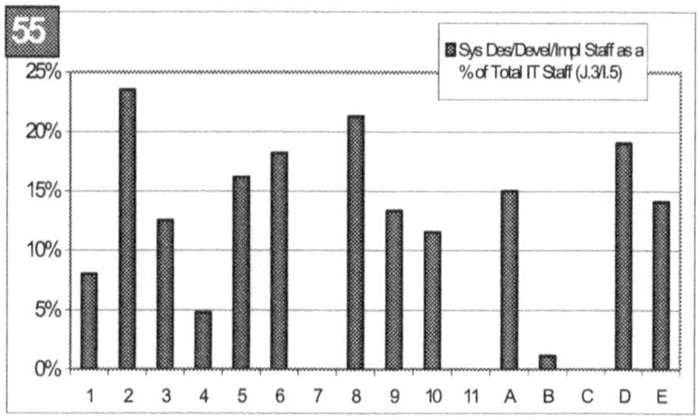
55
Sys Des/Devel/Impl Staff as a % of Total IT Staff (J.3/I.5)

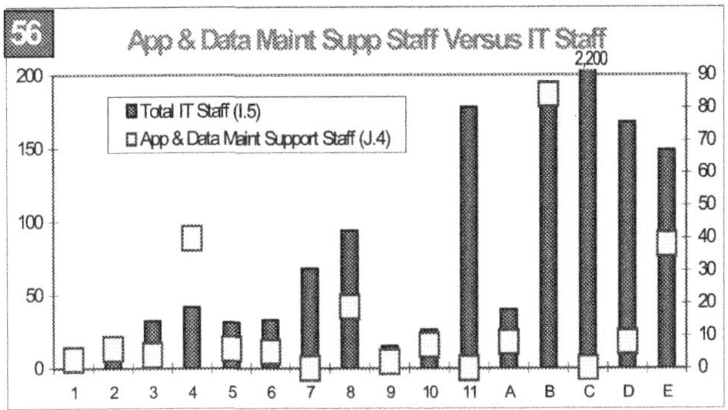
56
App & Data Maint Supp Staff Versus IT Staff
Total IT Staff (I.5)
App & Data Maint Support Staff (J.4)
2,200

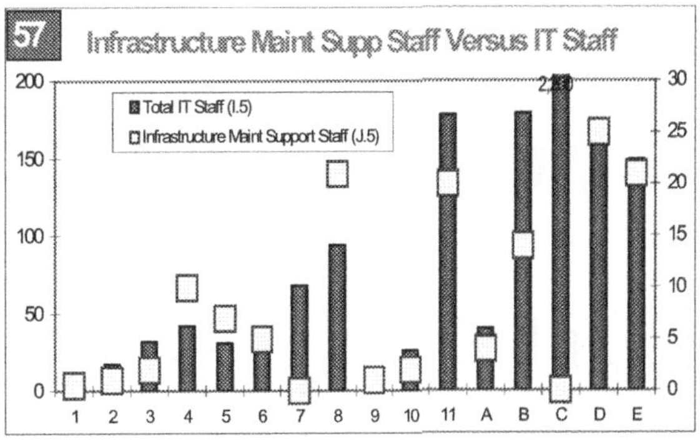
57
Infrastructure Maint Supp Staff Versus IT Staff
Total IT Staff (I.5)
Infrastructure Maint Support Staff (J.5)

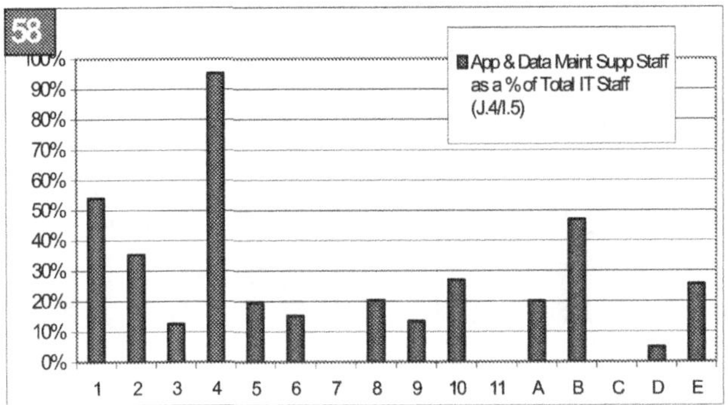
58
App & Data Maint Supp Staff as a % of Total IT Staff (J.4/I.5)

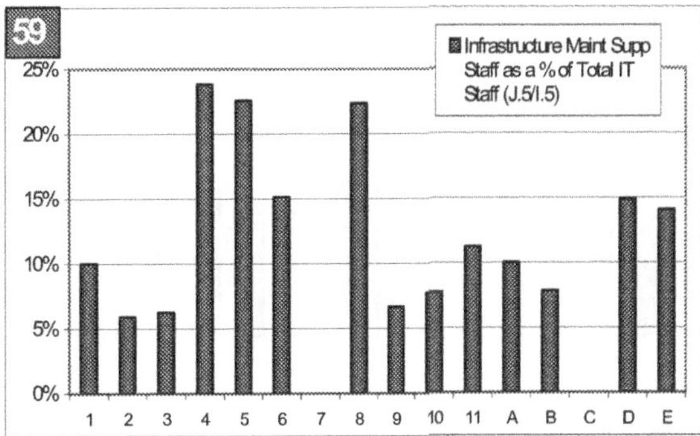
59
Infrastructure Maint Supp Staff as a % of Total IT Staff (J.5/I.5)

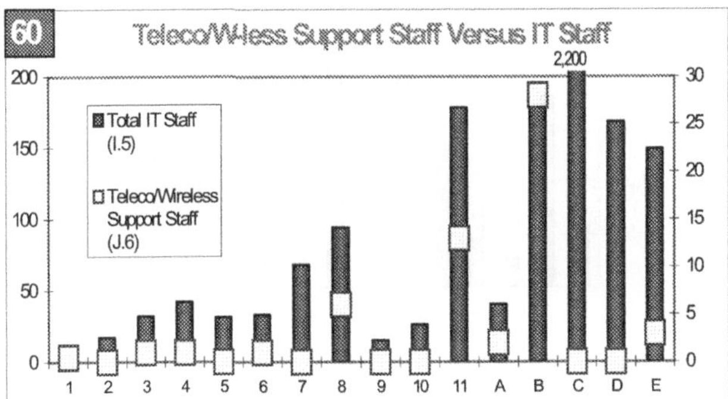
60
Teleco/W-less Support Staff Versus IT Staff
Total IT Staff (I.5)
Teleco/Wireless Support Staff (J.6)
2,200

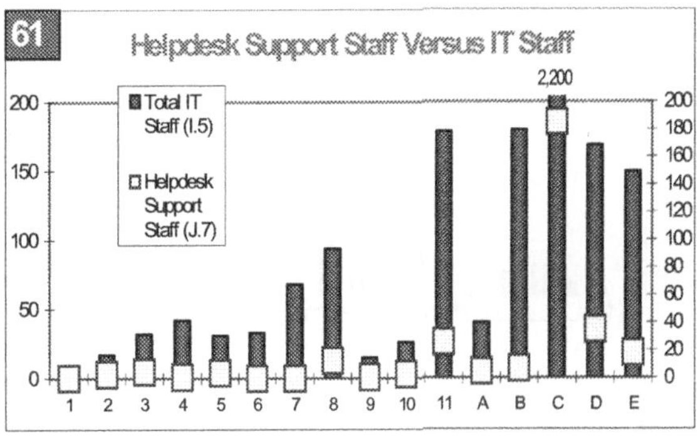
61
Helpdesk Support Staff Versus IT Staff
Total IT Staff (I.5)
Helpdesk Support Staff (J.7)
2,200

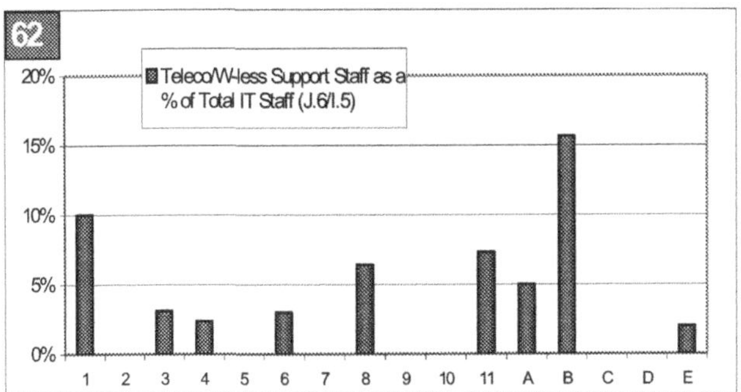
62
Teleco/W-less Support Staff as a % of Total IT Staff (J.6/I.5)

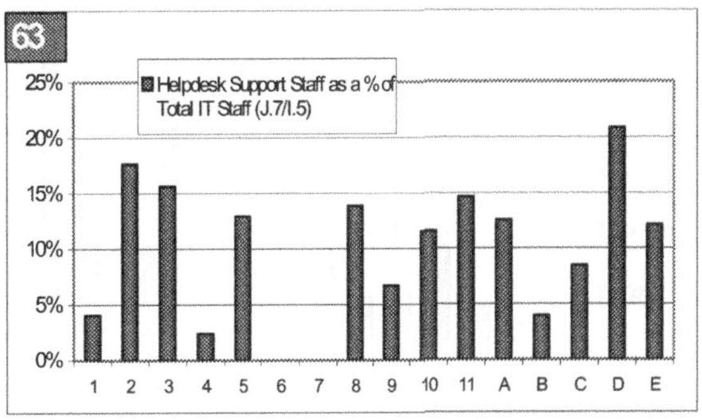
63
Helpdesk Support Staff as a % of Total IT Staff (J.7/I.5)

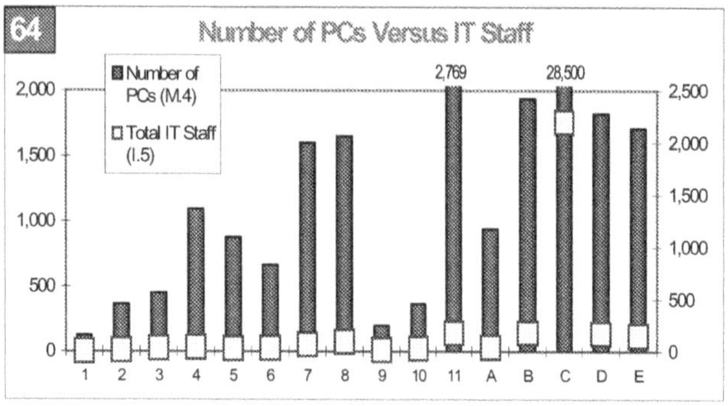
64
Number of PCs Versus IT Staff
Number of PCs (M.4)
Total IT Staff (I.5)
2,769
28,500

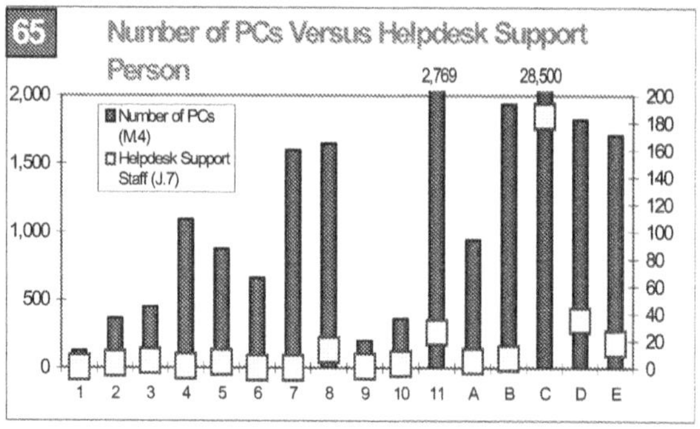
65
Number of PCs Versus Helpdesk Support Person
Number of PCs (M.4)
Helpdesk Support Staff (J.7)
2,769
28,500

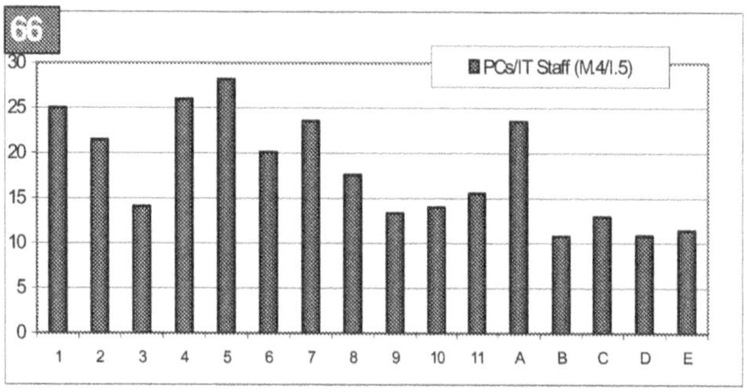
66
PCs/IT Staff (M.4/I.5)

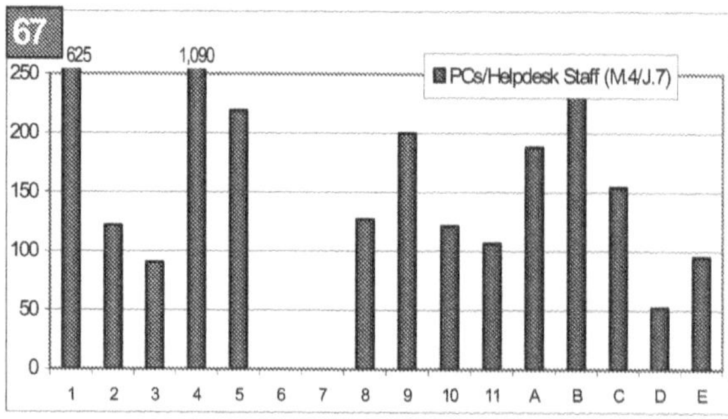
67
625
1,090
PCs/Helpdesk Staff (M.4/J.7)

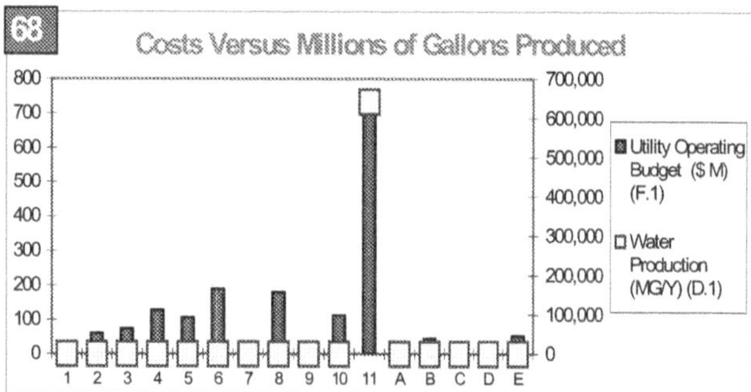
68
Costs Versus Millions of Gallons Produced
Utility Operating Budget ($ M) (F.1)
Water Production (MG/Y) (D.1)

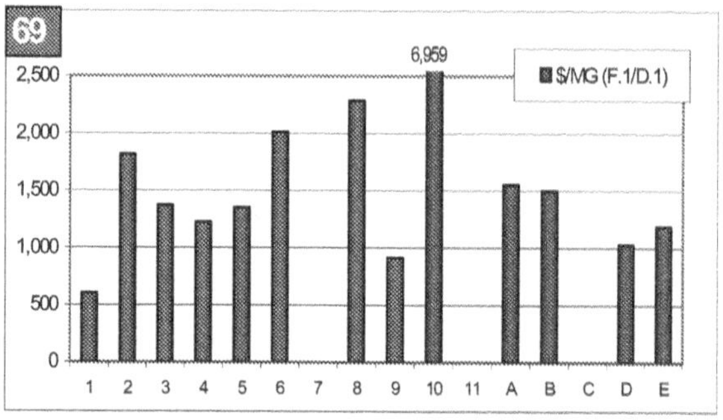
69
6,959
$/MG (F.1/D.1)

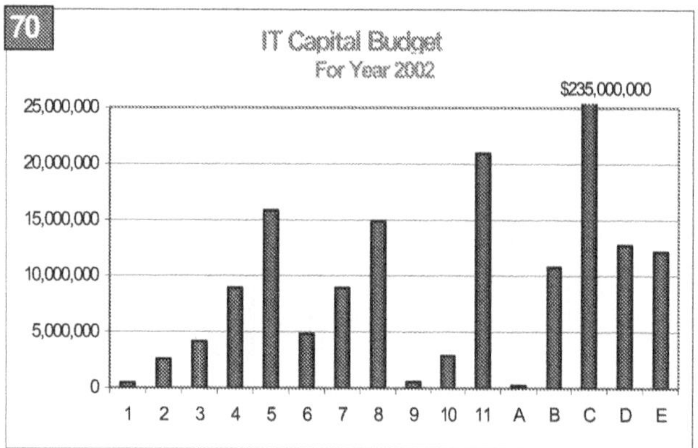
70
IT Capital Budget
For Year 2002
$235,000,000

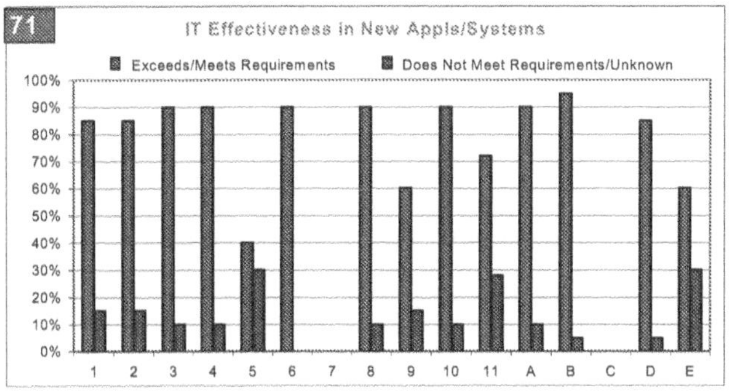


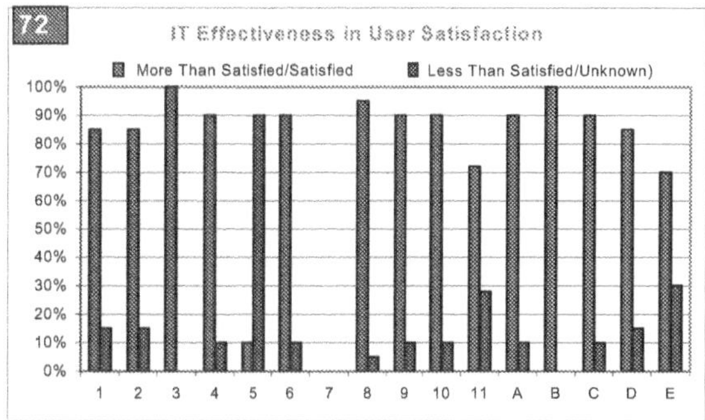


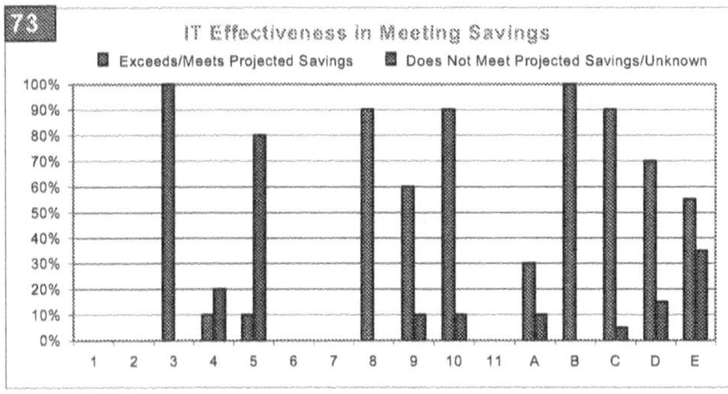


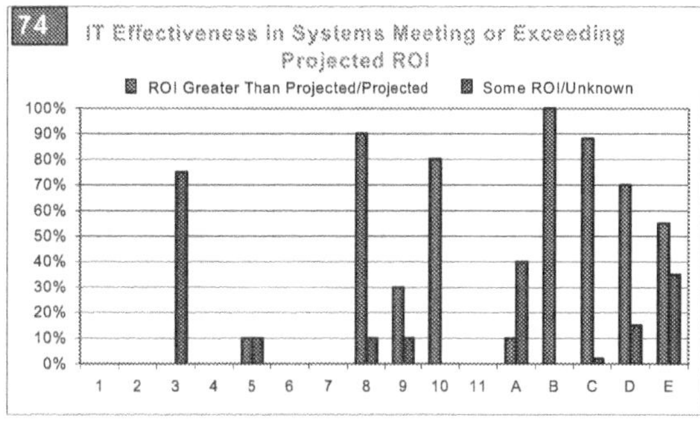


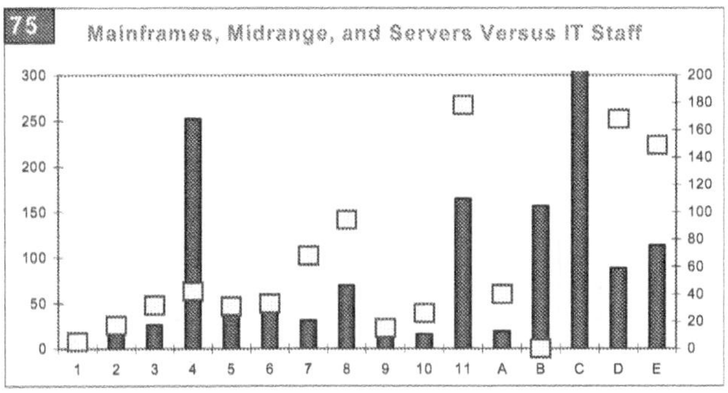


## 78 IT Staff Skills Balance

SH=Short Supply
B=Balanced
S=Surplus

| | SH=Short Supply; B=Balanced; S=Surplus | 1 | | | 2 | | | 3 | | | 4 | | | 5 | | | 6 | | | 7 | | | 8 | | | 9 | | | 10 | | | 11 | | | A | | | B | | | C | | | D | | | E | | |
|---|---|---|---|---|---|---|---|---|---|---|---|---|---|---|---|---|---|---|---|---|---|---|---|---|---|---|---|---|---|---|---|---|---|---|---|---|---|---|---|---|---|---|---|---|---|---|---|---|---|
| | | SH | B | S | SH | B | S | SH | B | S | SH | B | S | SH | B | S | SH | B | S | SH | B | S | SH | B | S | SH | B | S | SH | B | S | SH | B | S | SH | B | S | SH | B | S | SH | B | S | SH | B | S | SH | B | S |
| 1 | General Management and Planning | ■ | | | | ■ | | | ■ | | ■ | | | | ■ | | | ■ | | ■ | | | | ■ | | | ■ | | ■ | | | | ■ | | ■ | | | | ■ | | ■ | | | | ■ | | | ■ | |
| 2 | Business Application Analysis | | ■ | | ■ | | | | ■ | | ■ | | | ■ | | | ■ | | | ■ | | | ■ | | | | ■ | | ■ | | | | ■ | | ■ | | | | ■ | | ■ | | | ■ | | | ■ | | |
| 3 | System Design, Development, Implementation | ■ | | | ■ | | | ■ | | | ■ | | | ■ | | | ■ | | | ■ | | | ■ | | | ■ | | | ■ | | | | ■ | | | ■ | | | ■ | | ■ | | | | ■ | | ■ | | |
| 4 | Application and Data Maintenance/Support | | ■ | | | ■ | | | ■ | | ■ | | | ■ | | | ■ | | | | ■ | | | ■ | | | ■ | | ■ | | | | ■ | | | | ■ | | ■ | | | ■ | | ■ | | | ■ | | |
| 5 | Infrastructure Maintenance/Support (Networks, Servers, PCs) | ■ | | | ■ | | | | ■ | | | ■ | | | ■ | | ■ | | | | ■ | | | ■ | | | ■ | | ■ | | | | ■ | | | ■ | | | ■ | | | ■ | | | ■ | | | ■ | |
| 6 | Telecoms/Wireless Support | | ■ | | ■ | | | ■ | | | | ■ | | | | | | ■ | | ■ | | | ■ | | | ■ | | | | | | | ■ | | | ■ | | | ■ | | | ■ | | | ■ | | | ■ | |
| 7 | Help Desk/Desktop Support | ■ | | | ■ | | | | ■ | | ■ | | | | ■ | | ■ | | | | ■ | | | ■ | | | ■ | | | ■ | | | | ■ | | ■ | | | ■ | | | ■ | | | ■ | | | ■ | |
| 8 | Other (specify): Management | | | | | | | | | | | | | | | | | | | | | | | | | | | | | | | | | | | | | | | | | | | | ■ | | | | |
| 8 | Other (specify): GIS | | | | | | | | | | | | | | | | | | | | | | | | | | | | | | | ■ | | | | | | | | | | | | | | | | | |
| 8 | Other (specify): Management Analyst | | | | ■ | | | | | | | | | | | | | | | | | | | | | | | | | | | | | | | | | | | | | | | | | | | | |
| 8 | Other (specify): Computer Room Operations | | | | | | | | | | | | | | | | | | | | | | | | | | | | | | | | | | | | | | ■ | | | | | | | | | | |

The following are individual charts representing responses for ***all*** practice questions in each focus area by participating utilities. A scale of 1-6 is from worst to best, with a rating of 6 representing leading practices.

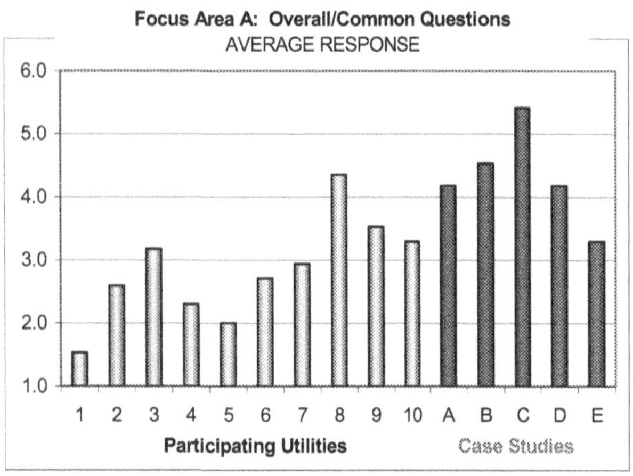


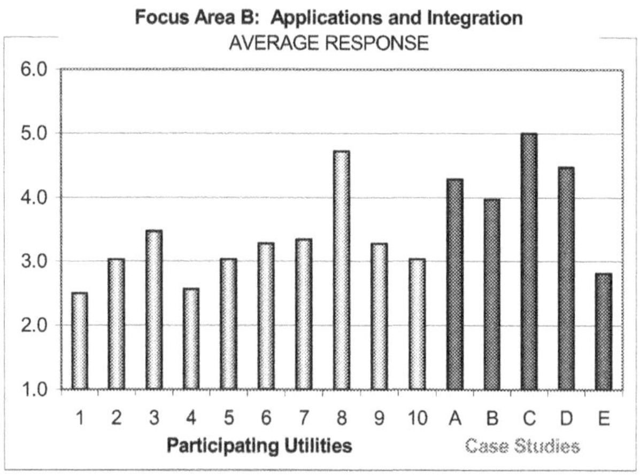


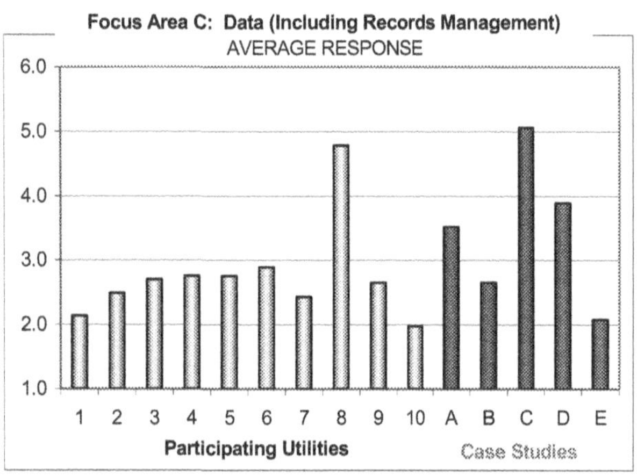


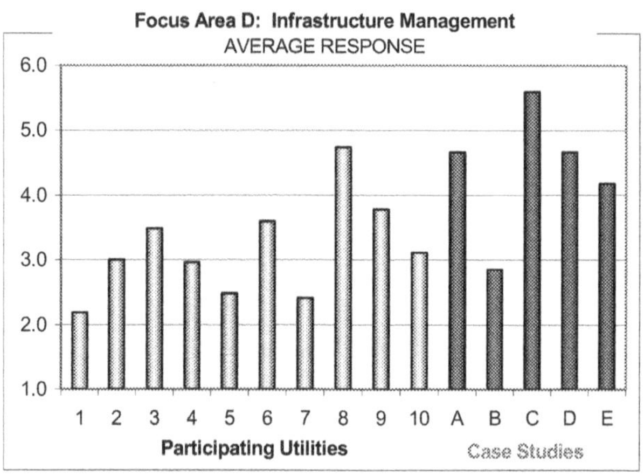

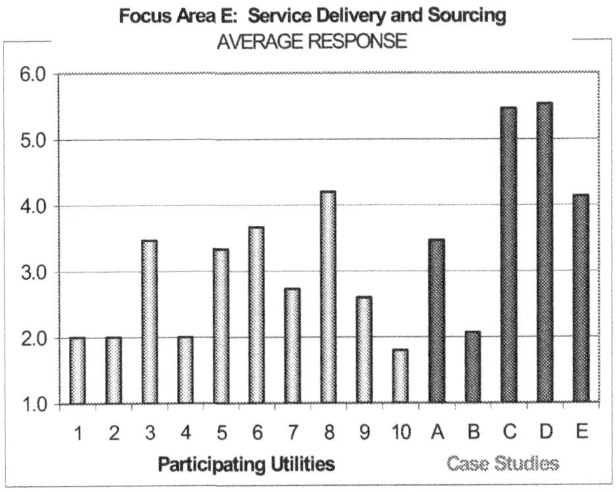
Focus Area E: Service Delivery and Sourcing
AVERAGE RESPONSE
6.0
5.0
4.0
3.0
2.0
1.0
1 2 3 4 5 6 7 8 9 10 A B C D E
Participating Utilities
Case Studies

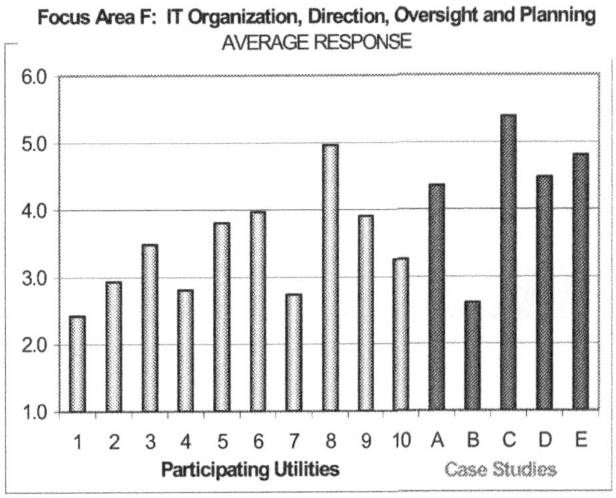
Focus Area F: IT Organization, Direction, Oversight and Planning
AVERAGE RESPONSE
6.0
5.0
4.0
3.0
2.0
1.0
1 2 3 4 5 6 7 8 9 10 A B C D E
Participating Utilities
Case Studies

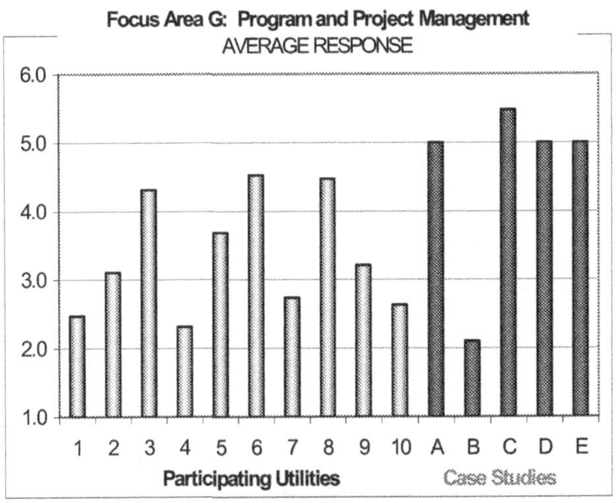
Focus Area G: Program and Project Management
AVERAGE RESPONSE
6.0
5.0
4.0
3.0
2.0
1.0
1 2 3 4 5 6 7 8 9 10 A B C D E
Participating Utilities
Case Studies

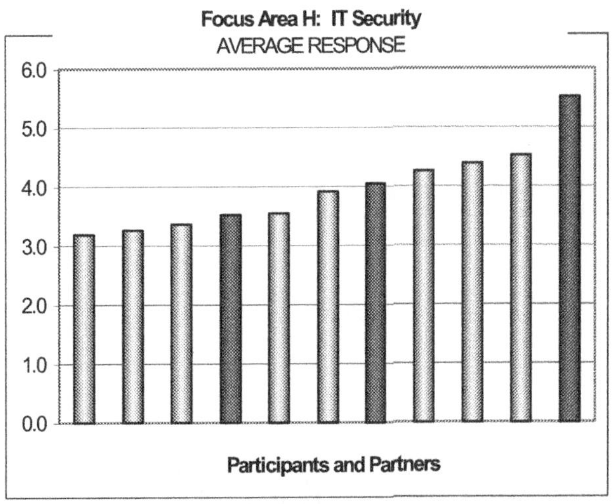
Focus Area H: IT Security
AVERAGE RESPONSE
6.0
5.0
4.0
3.0
2.0
1.0
0.0
Participants and Partners

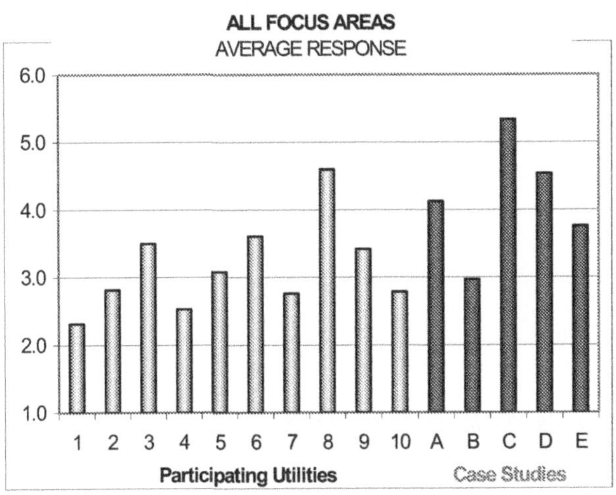
ALL FOCUS AREAS
AVERAGE RESPONSE
6.0
5.0
4.0
3.0
2.0
1.0
1 2 3 4 5 6 7 8 9 10 A B C D E
Participating Utilities
Case Studies

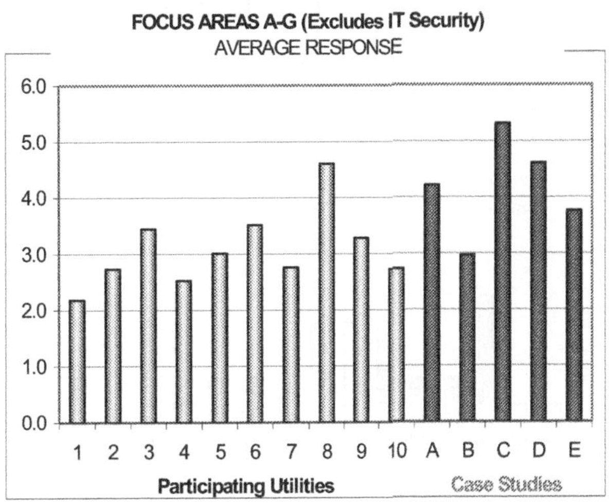
FOCUS AREAS A-G (Excludes IT Security)
AVERAGE RESPONSE
6.0
5.0
4.0
3.0
2.0
1.0
0.0
1 2 3 4 5 6 7 8 9 10 A B C D E
Participating Utilities
Case Studies

## PRACTICE QUESTIONS RESPONSES

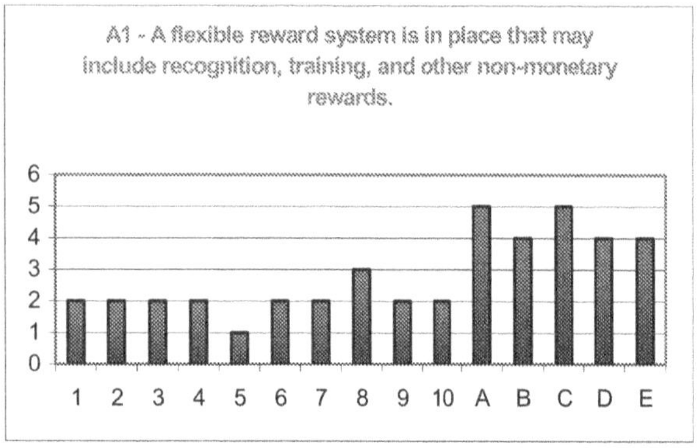


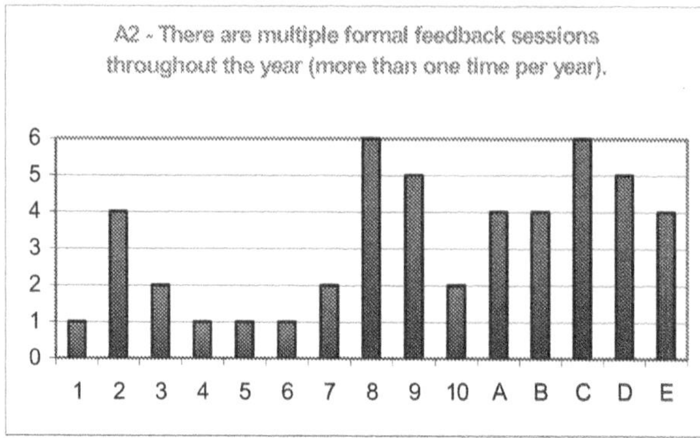


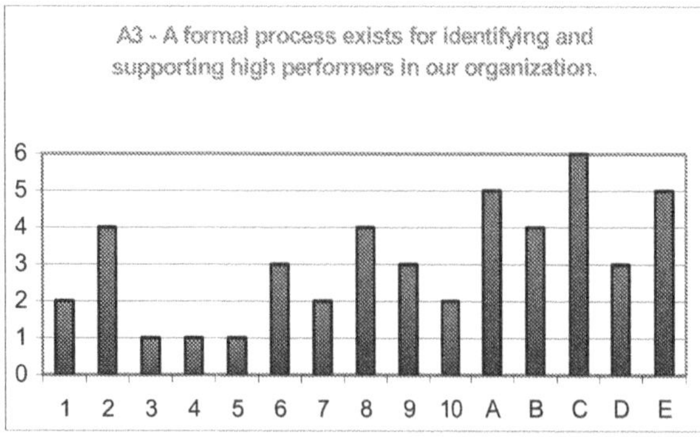


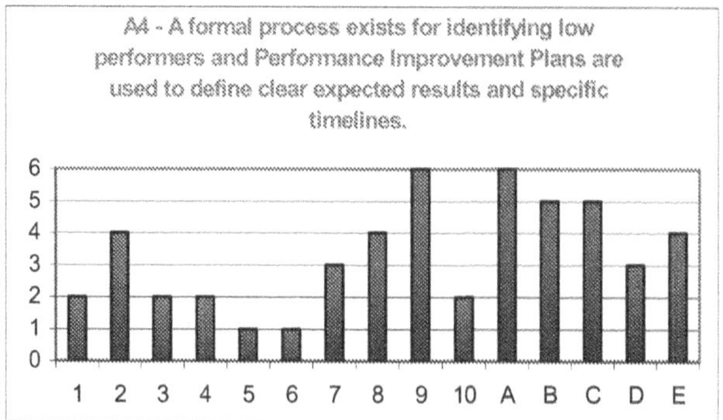


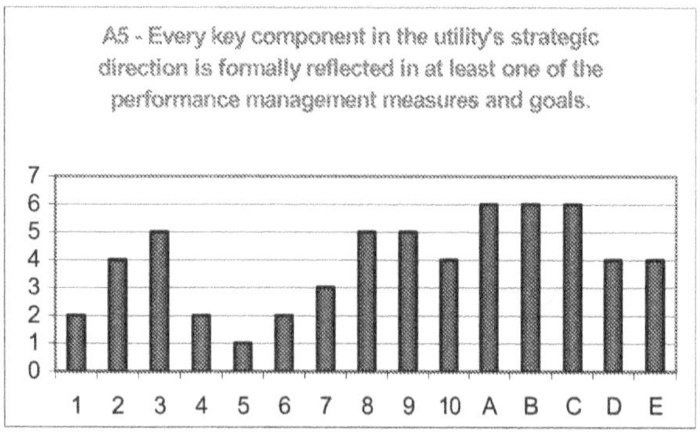


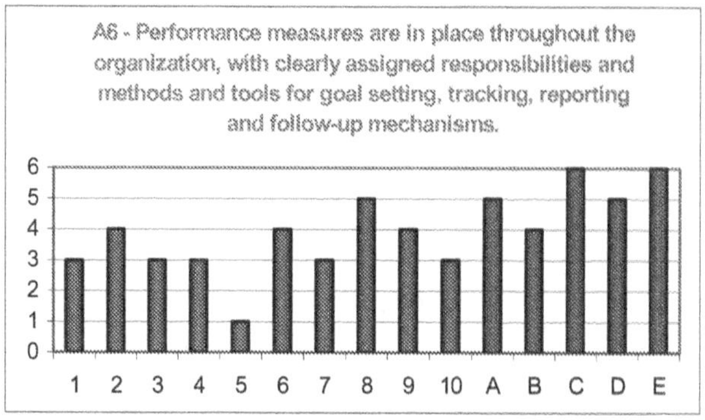


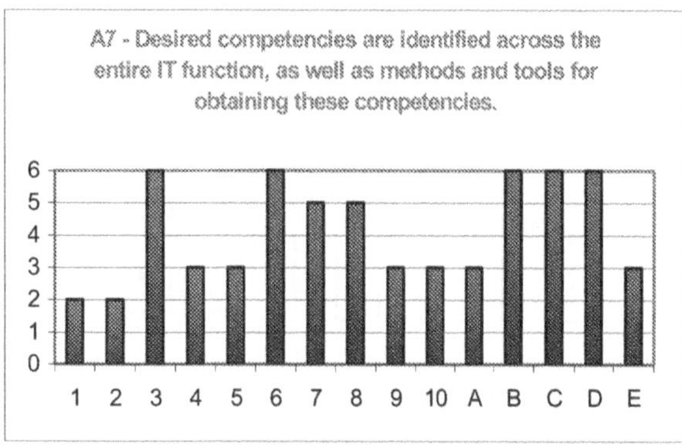


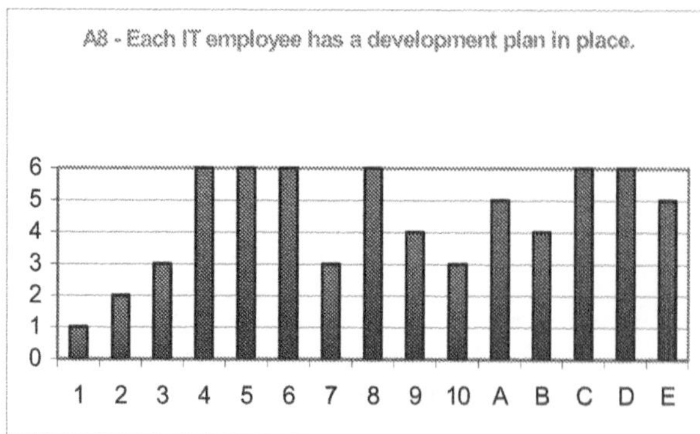

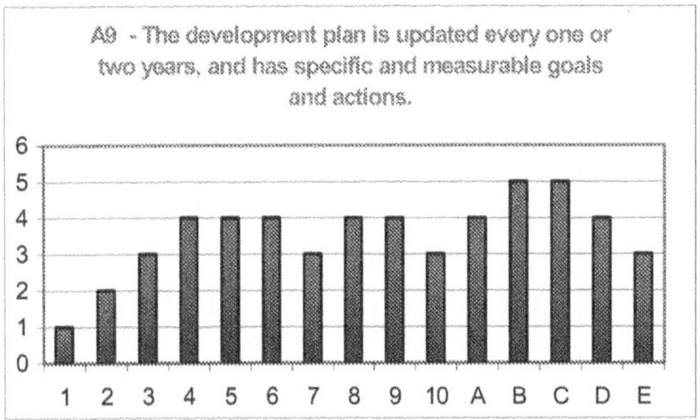
A9 - The development plan is updated every one or two years, and has specific and measurable goals and actions.
0 1 2 3 4 5 6
1 2 3 4 5 6 7 8 9 10 A B C D E

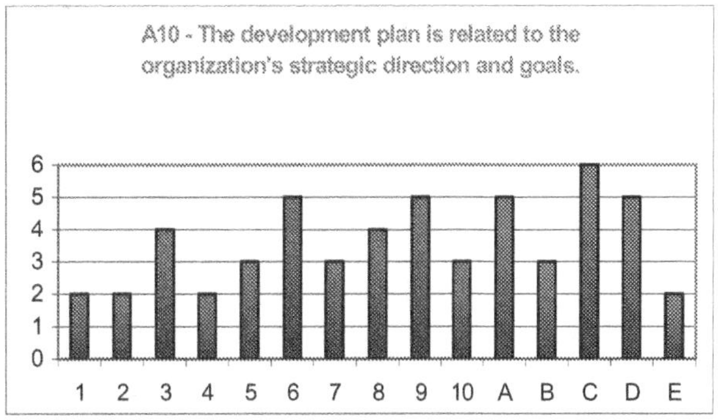
A10 - The development plan is related to the organization's strategic direction and goals.
0 1 2 3 4 5 6
1 2 3 4 5 6 7 8 9 10 A B C D E

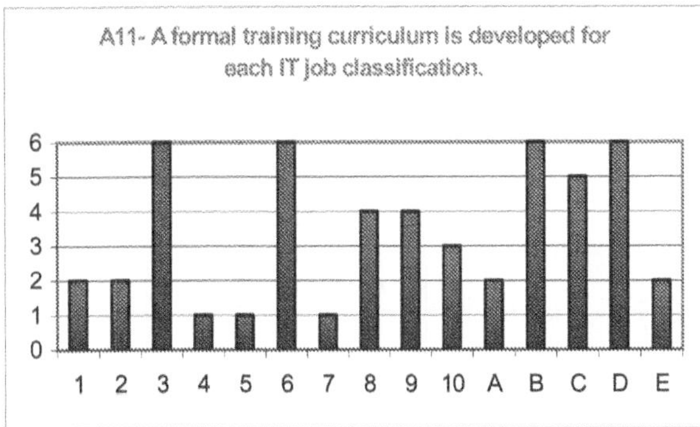
A11- A formal training curriculum is developed for each IT job classification.
0 1 2 3 4 5 6
1 2 3 4 5 6 7 8 9 10 A B C D E

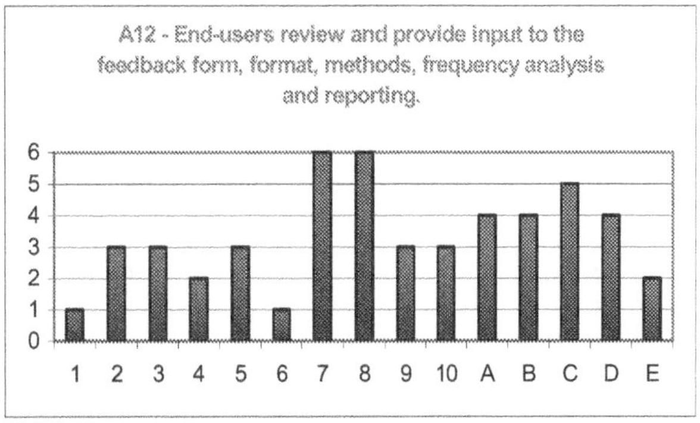
A12 - End-users review and provide input to the feedback form, format, methods, frequency analysis and reporting.
0 1 2 3 4 5 6
1 2 3 4 5 6 7 8 9 10 A B C D E

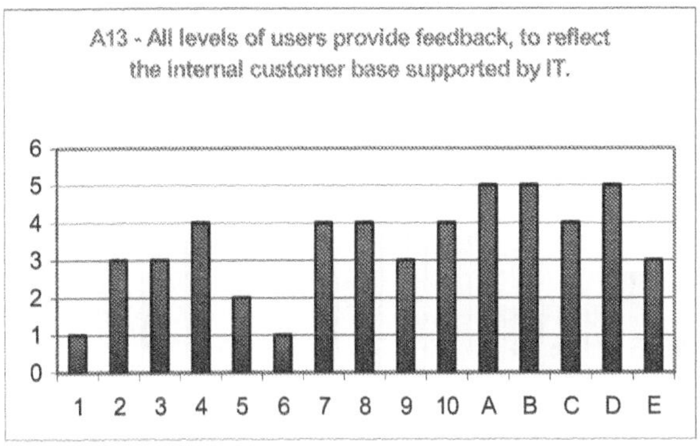
A13 - All levels of users provide feedback, to reflect the internal customer base supported by IT.
0 1 2 3 4 5 6
1 2 3 4 5 6 7 8 9 10 A B C D E

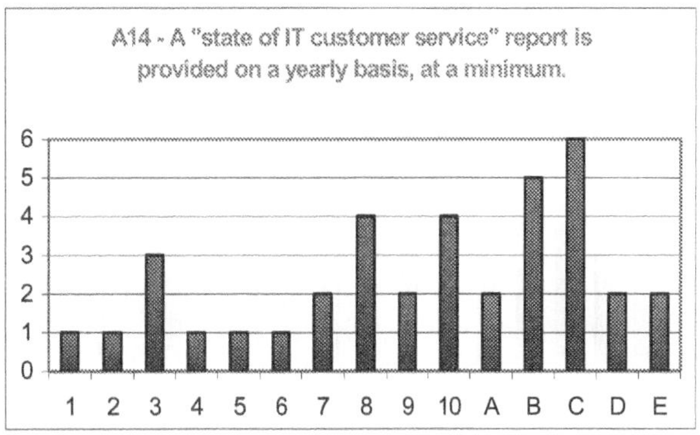
A14 - A "state of IT customer service" report is provided on a yearly basis, at a minimum.
0 1 2 3 4 5 6
1 2 3 4 5 6 7 8 9 10 A B C D E

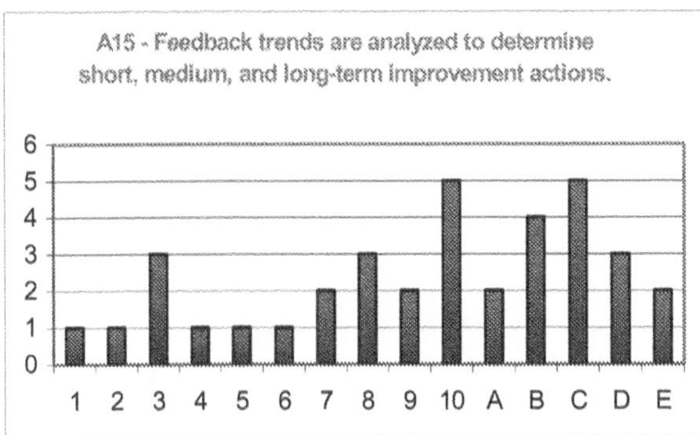
A15 - Feedback trends are analyzed to determine short, medium, and long-term improvement actions.
0 1 2 3 4 5 6
1 2 3 4 5 6 7 8 9 10 A B C D E

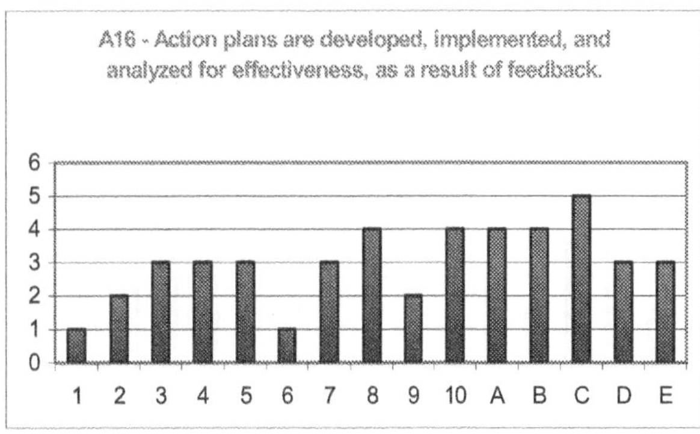
A16 - Action plans are developed, implemented, and analyzed for effectiveness, as a result of feedback.
0 1 2 3 4 5 6
1 2 3 4 5 6 7 8 9 10 A B C D E

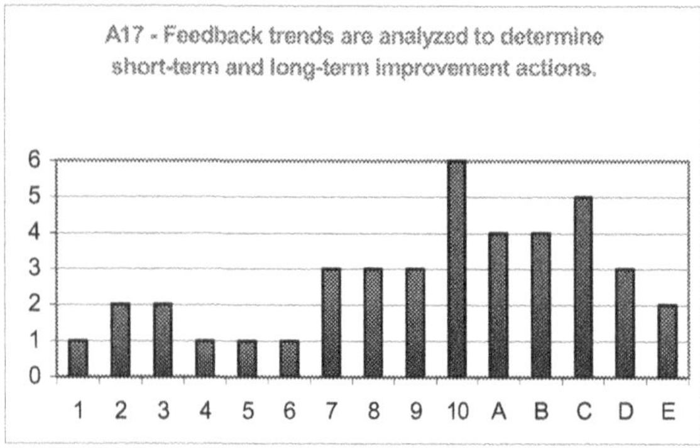
A17 - Feedback trends are analyzed to determine short-term and long-term improvement actions.
0 1 2 3 4 5 6
1 2 3 4 5 6 7 8 9 10 A B C D E

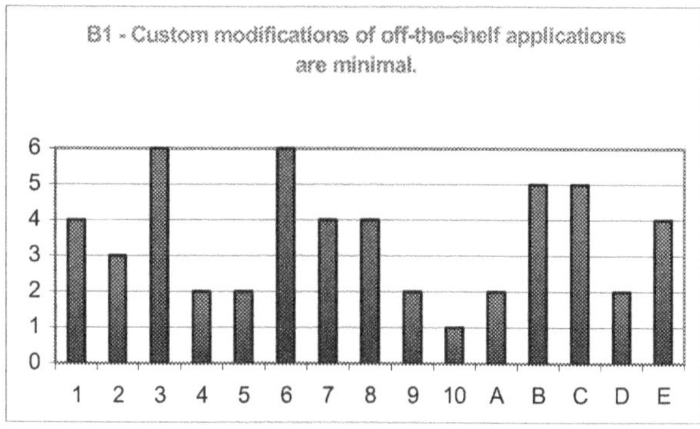
B1 - Custom modifications of off-the-shelf applications are minimal.
0 1 2 3 4 5 6
1 2 3 4 5 6 7 8 9 10 A B C D E

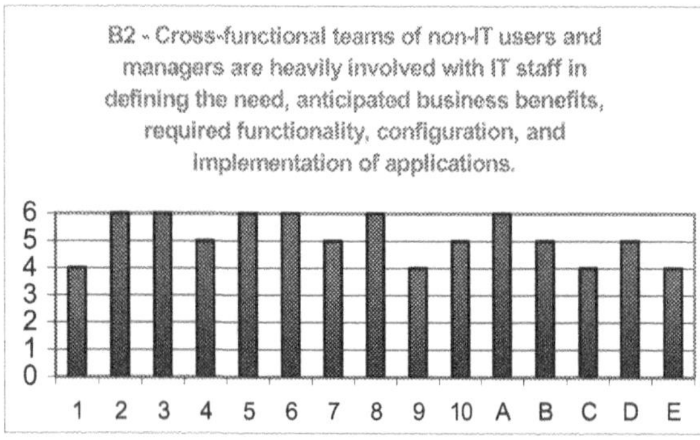
B2 - Cross-functional teams of non-IT users and managers are heavily involved with IT staff in defining the need, anticipated business benefits, required functionality, configuration, and implementation of applications.
0 1 2 3 4 5 6
1 2 3 4 5 6 7 8 9 10 A B C D E

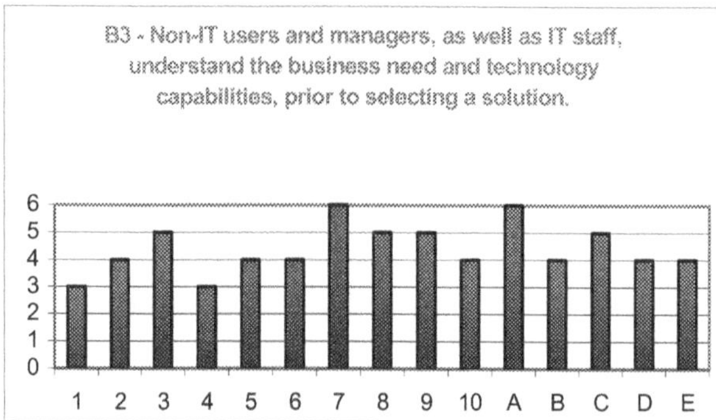
B3 - Non-IT users and managers, as well as IT staff, understand the business need and technology capabilities, prior to selecting a solution.
0 1 2 3 4 5 6
1 2 3 4 5 6 7 8 9 10 A B C D E

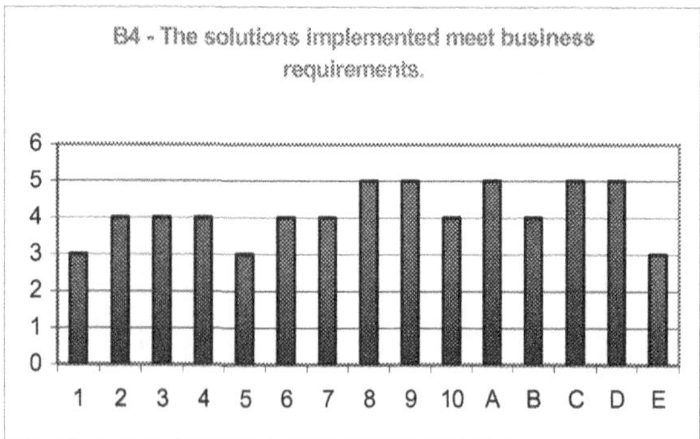
B4 - The solutions implemented meet business requirements.
0 1 2 3 4 5 6
1 2 3 4 5 6 7 8 9 10 A B C D E

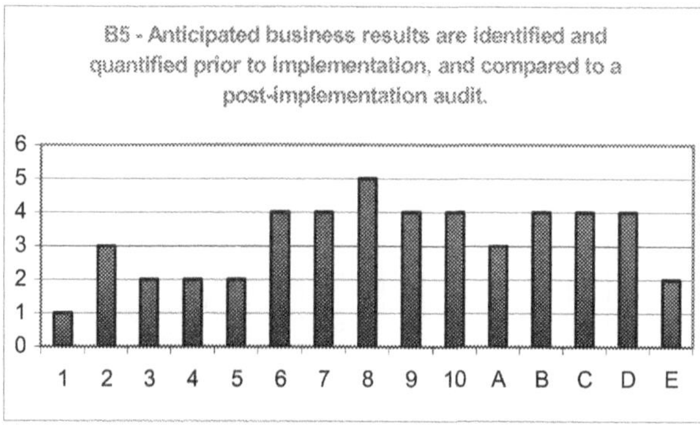
B5 - Anticipated business results are identified and quantified prior to implementation, and compared to a post-implementation audit.
0 1 2 3 4 5 6
1 2 3 4 5 6 7 8 9 10 A B C D E

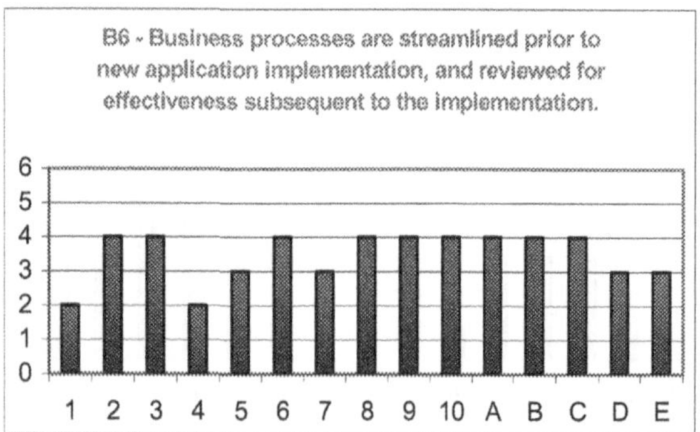
B6 - Business processes are streamlined prior to new application implementation, and reviewed for effectiveness subsequent to the implementation.
0 1 2 3 4 5 6
1 2 3 4 5 6 7 8 9 10 A B C D E

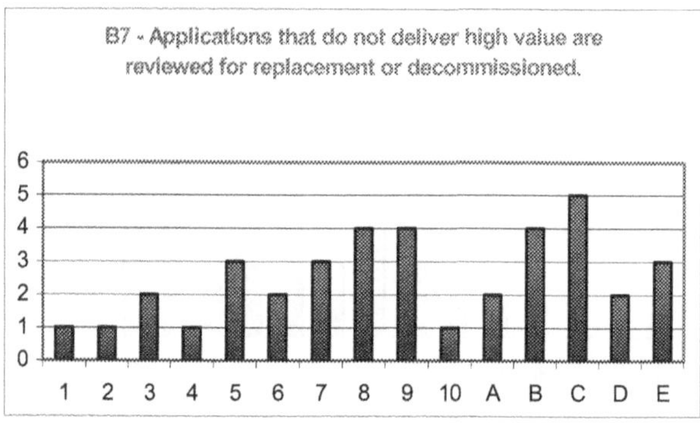
B7 - Applications that do not deliver high value are reviewed for replacement or decommissioned.
0 1 2 3 4 5 6
1 2 3 4 5 6 7 8 9 10 A B C D E

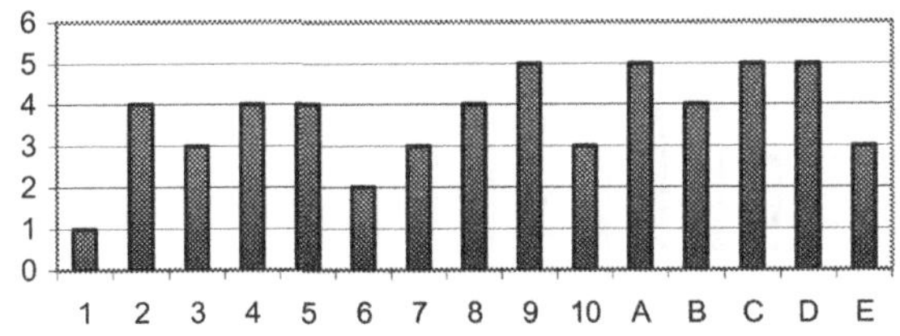

B8 - The cost (labor, maintenance contract, support tools) of maintaining applications are identified, monitored and tracked.

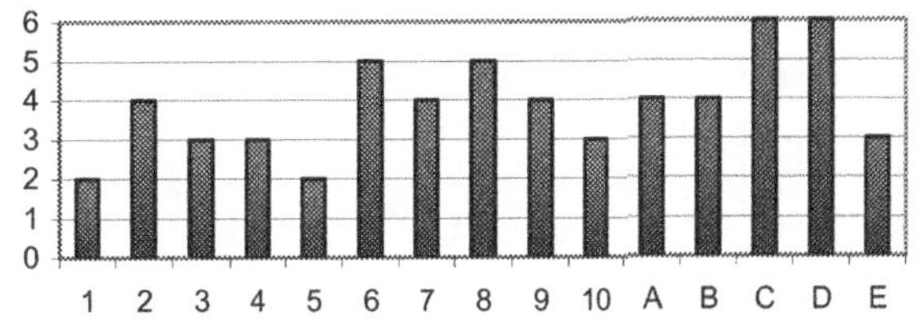

B9 - Requests for new functionality are reviewed to determine if an existing application can meet the requirements.

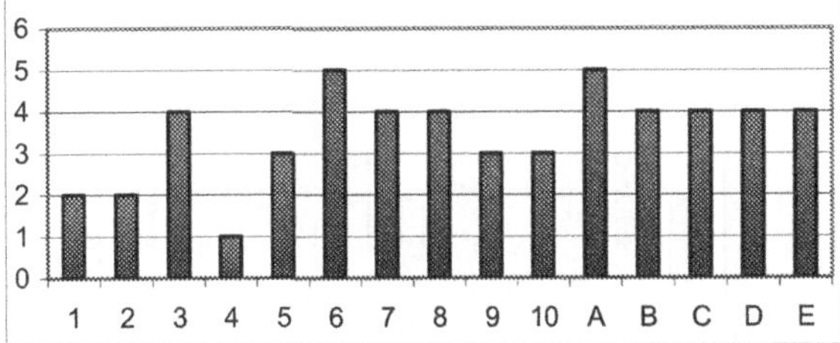

B10 - Requests for new functionality are reviewed to determine if they support the strategic business plan.

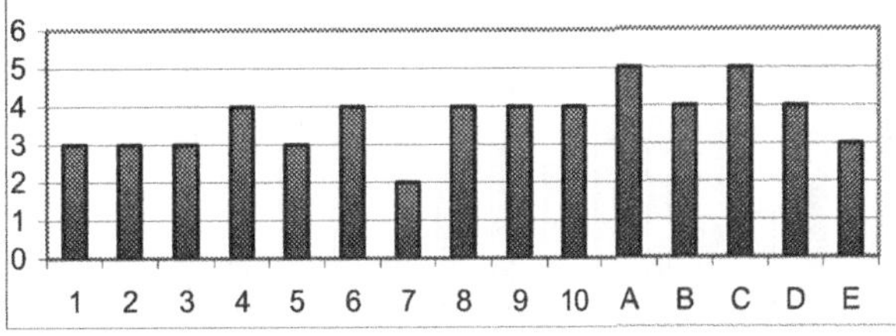

B11 - New applications are considered in the context of existing applications, with the result that existing applications may be decommissioned.

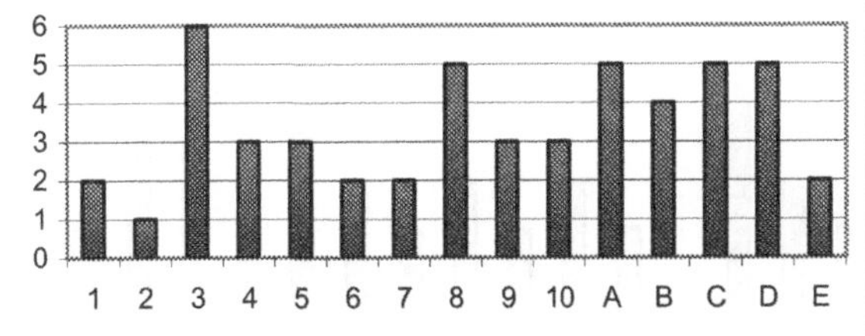

B12 - A suite of tools and associated standards and guidelines exists and is used for all application development projects.

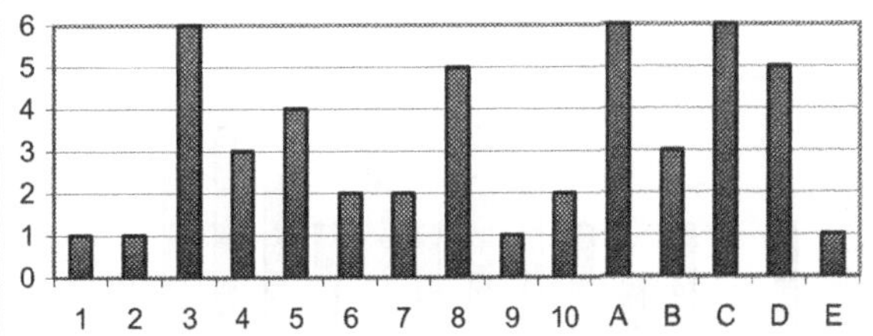

B13 - An application development methodology is documented and is current.

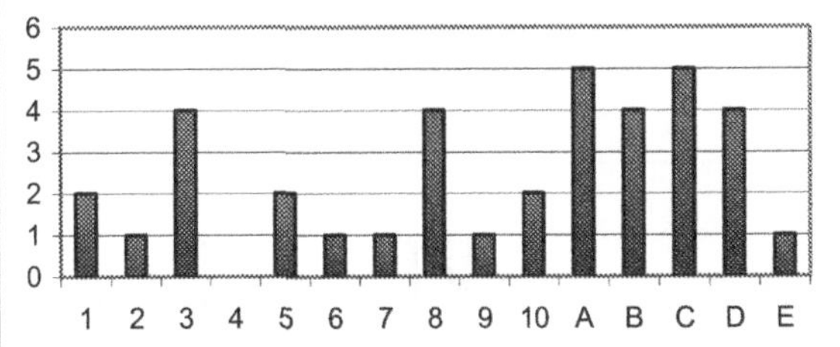

B14 - A formal process is in place to train application developers in the effective use of the development tools and methodologies.

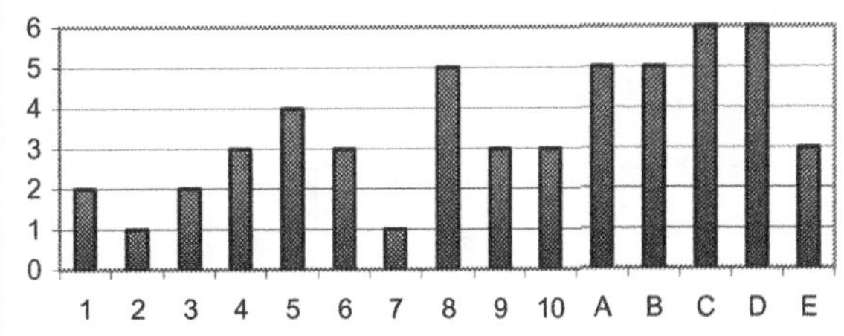

B15 - Costs to develop and enhance applications are tracked and formally reported.

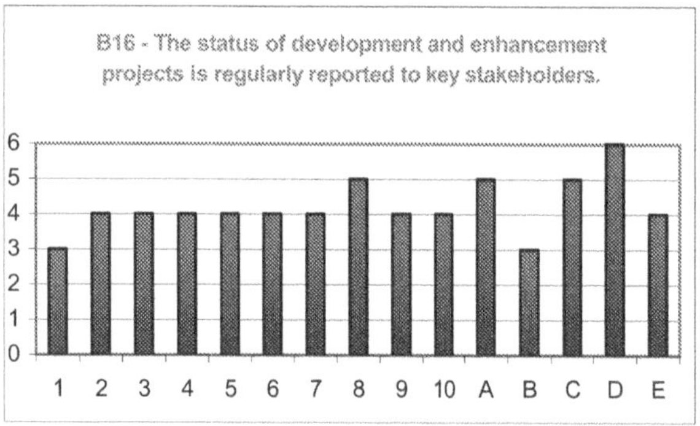
B16 - The status of development and enhancement
projects is regularly reported to key stakeholders.
0 1 2 3 4 5 6
1 2 3 4 5 6 7 8 9 10 A B C D E

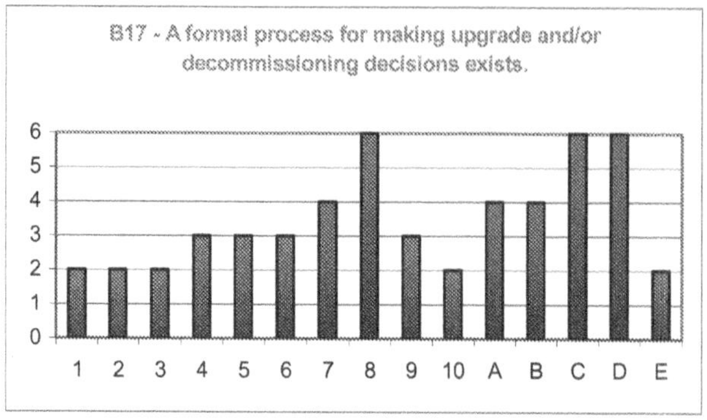
B17 - A formal process for making upgrade and/or
decommissioning decisions exists.
0 1 2 3 4 5 6
1 2 3 4 5 6 7 8 9 10 A B C D E

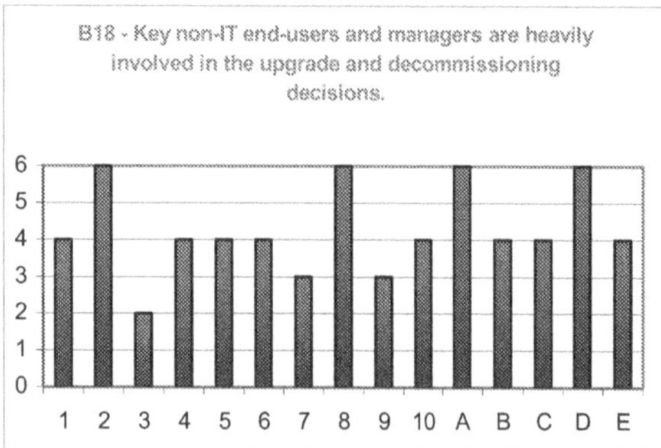
B18 - Key non-IT end-users and managers are heavily
involved in the upgrade and decommissioning
decisions.
0 1 2 3 4 5 6
1 2 3 4 5 6 7 8 9 10 A B C D E

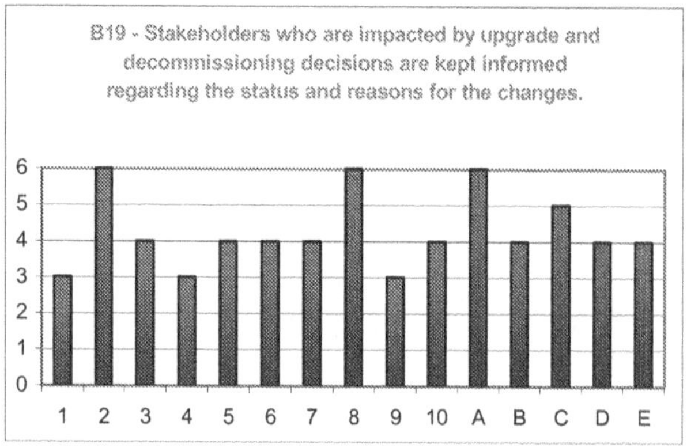
B19 - Stakeholders who are impacted by upgrade and
decommissioning decisions are kept informed
regarding the status and reasons for the changes.
0 1 2 3 4 5 6
1 2 3 4 5 6 7 8 9 10 A B C D E

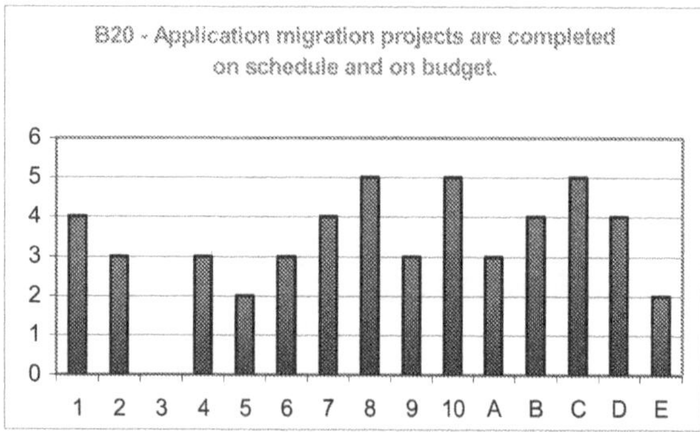
B20 - Application migration projects are completed
on schedule and on budget.
0 1 2 3 4 5 6
1 2 3 4 5 6 7 8 9 10 A B C D E

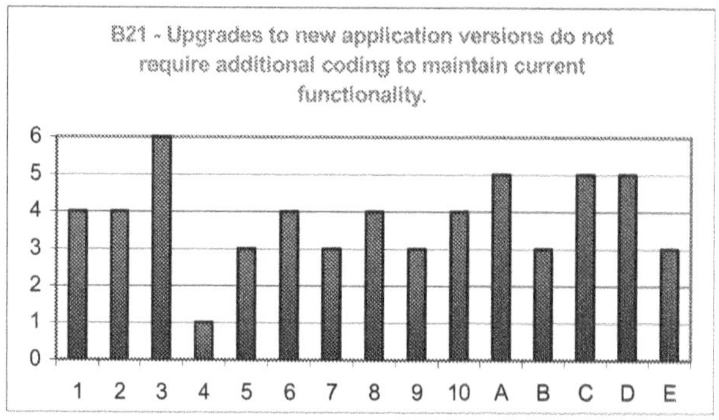
B21 - Upgrades to new application versions do not
require additional coding to maintain current
functionality.
0 1 2 3 4 5 6
1 2 3 4 5 6 7 8 9 10 A B C D E

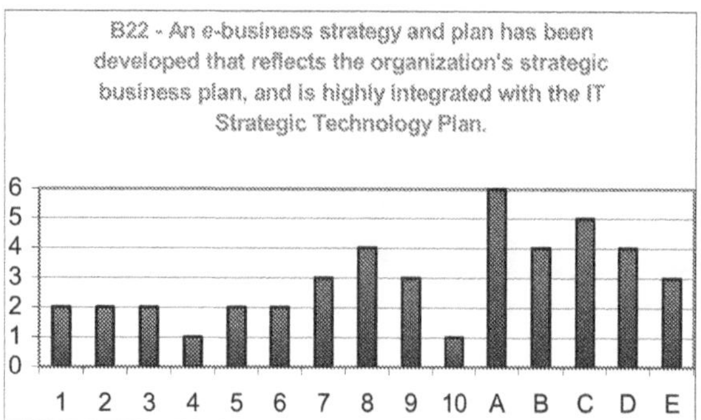
B22 - An e-business strategy and plan has been
developed that reflects the organization's strategic
business plan, and is highly integrated with the IT
Strategic Technology Plan.
0 1 2 3 4 5 6
1 2 3 4 5 6 7 8 9 10 A B C D E

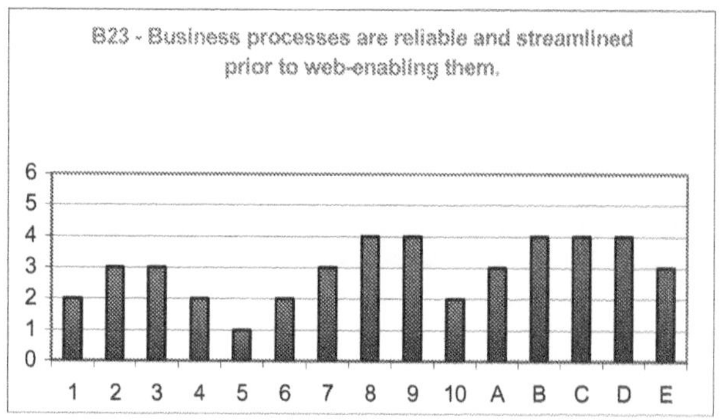
B23 - Business processes are reliable and streamlined
prior to web-enabling them.
0 1 2 3 4 5 6
1 2 3 4 5 6 7 8 9 10 A B C D E

B24 - End-users use (near)-real-time data, and to make decisions based upon that data. They trust the data in the system.

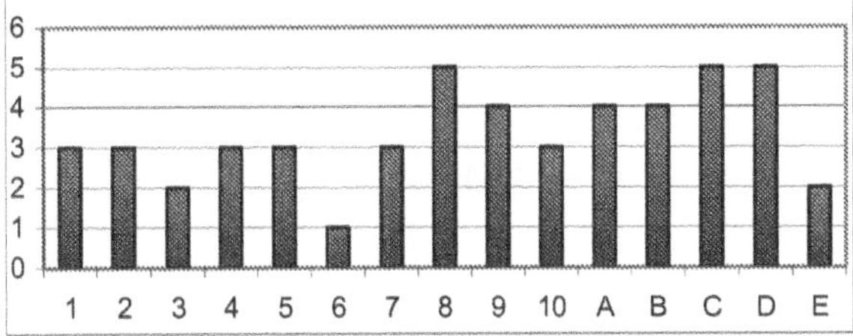

B25 - Clear and formal guidelines exist regarding what data may be published both within and outside the utility. Application configurations support these guidelines.

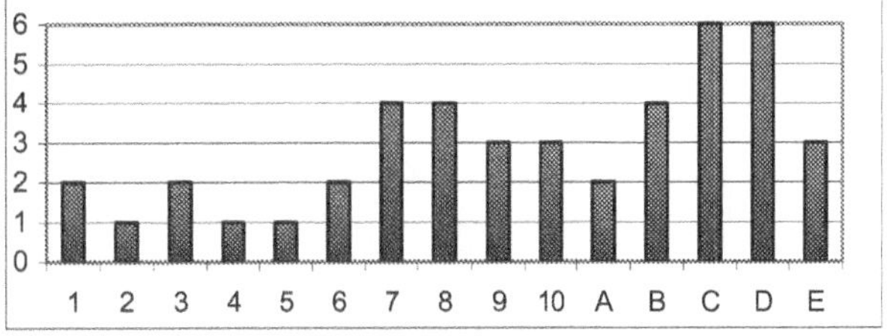

B26 - A system is in place to quickly and correctly address questions regarding application usage.

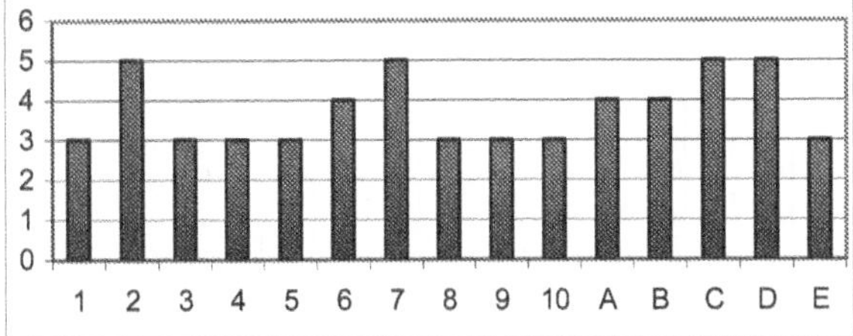

B27 - A formal process exists to ensure that applications sharing valuable utility data are prioritized for integration projects.

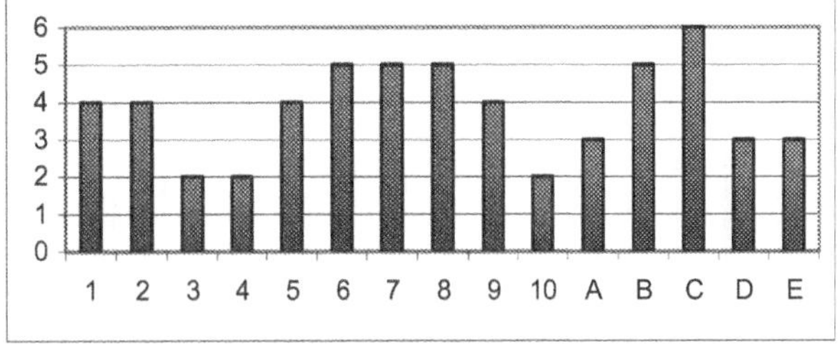

B28 - Applications providing easier integration methods are given preference when considering procuring or upgrading applications.

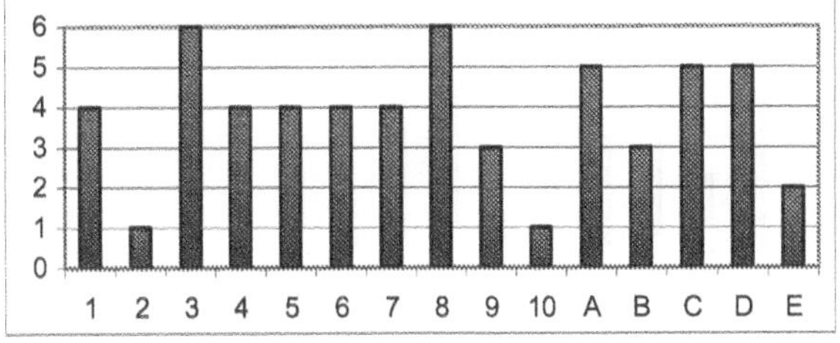

B29 - A formal process exists to ensure that vendor source code is not modified.

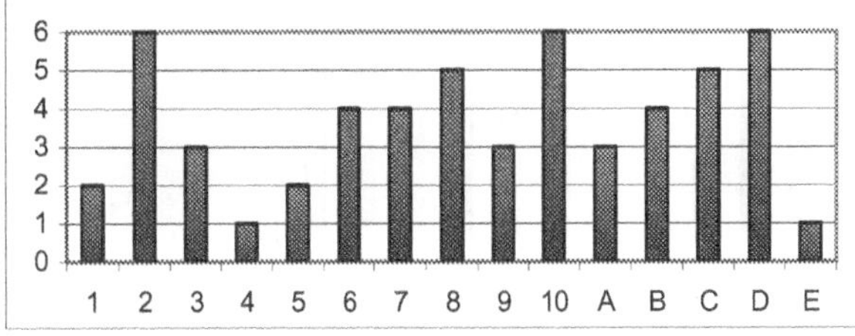

B30 - A business case analysis is conducted to compare the costs of integrating applications on an application-to-application-specific basis, to the cost of integrating applications on a one-to-many basis.

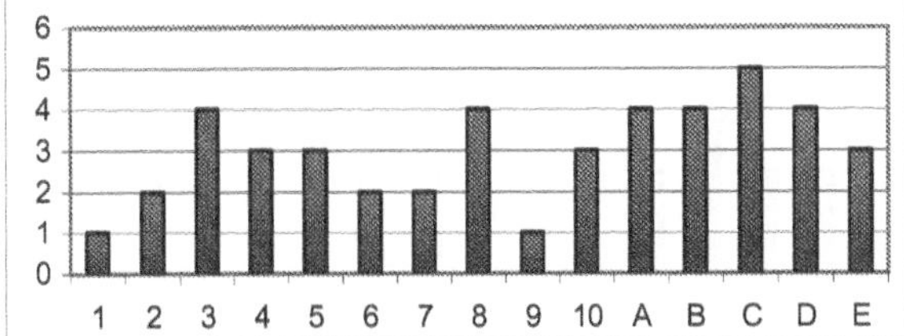

B31 - The cost of "bad data" is quantified for key business processes, and is factored into the integration business case.

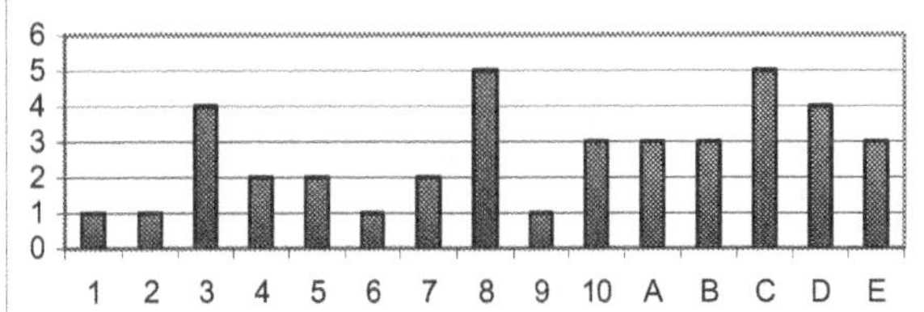

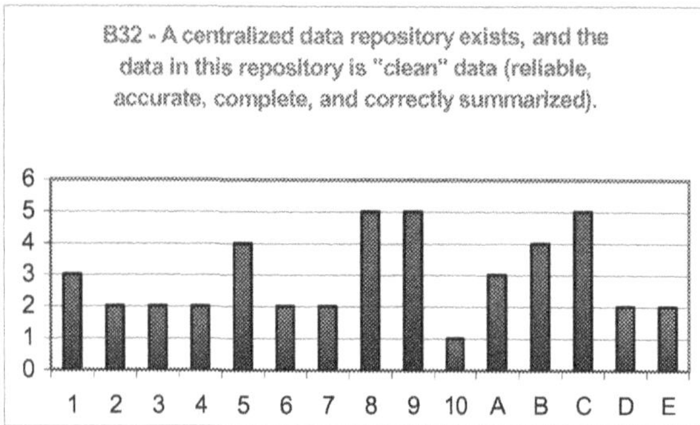
B32 - A centralized data repository exists, and the data in this repository is "clean" data (reliable, accurate, complete, and correctly summarized).
0 1 2 3 4 5 6
1 2 3 4 5 6 7 8 9 10 A B C D E

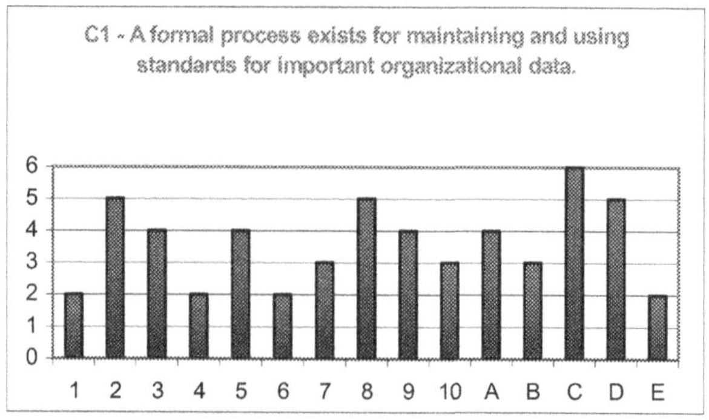
C1 - A formal process exists for maintaining and using standards for important organizational data.
0 1 2 3 4 5 6
1 2 3 4 5 6 7 8 9 10 A B C D E

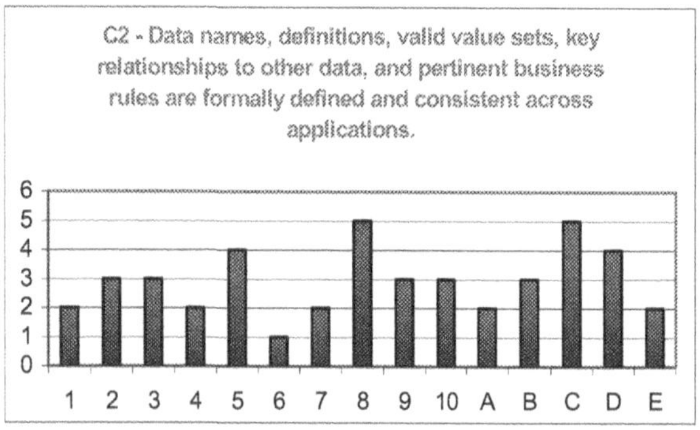
C2 - Data names, definitions, valid value sets, key relationships to other data, and pertinent business rules are formally defined and consistent across applications.
0 1 2 3 4 5 6
1 2 3 4 5 6 7 8 9 10 A B C D E

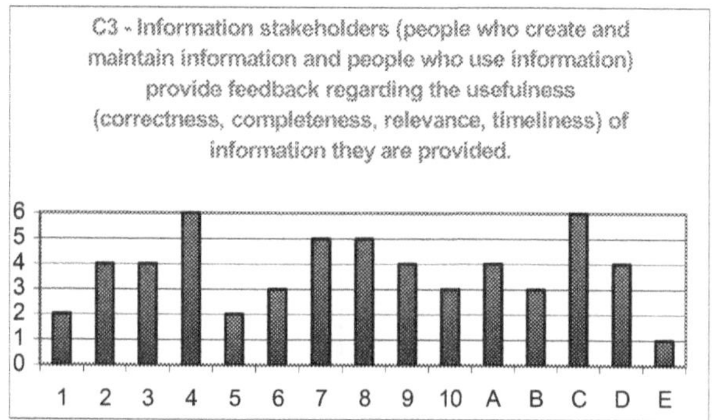
C3 - Information stakeholders (people who create and maintain information and people who use information) provide feedback regarding the usefulness (correctness, completeness, relevance, timeliness) of information they are provided.
0 1 2 3 4 5 6
1 2 3 4 5 6 7 8 9 10 A B C D E

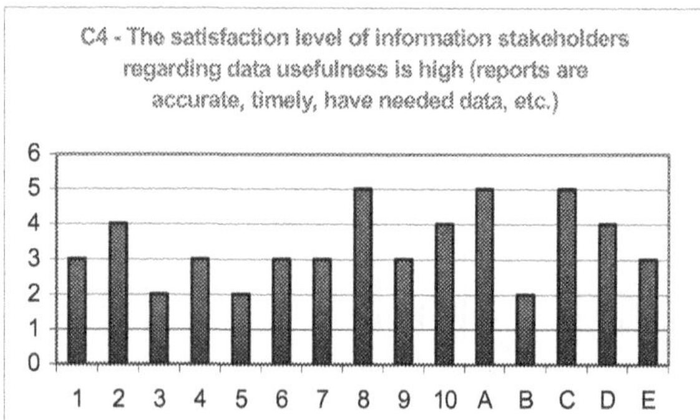
C4 - The satisfaction level of information stakeholders regarding data usefulness is high (reports are accurate, timely, have needed data, etc.)
0 1 2 3 4 5 6
1 2 3 4 5 6 7 8 9 10 A B C D E

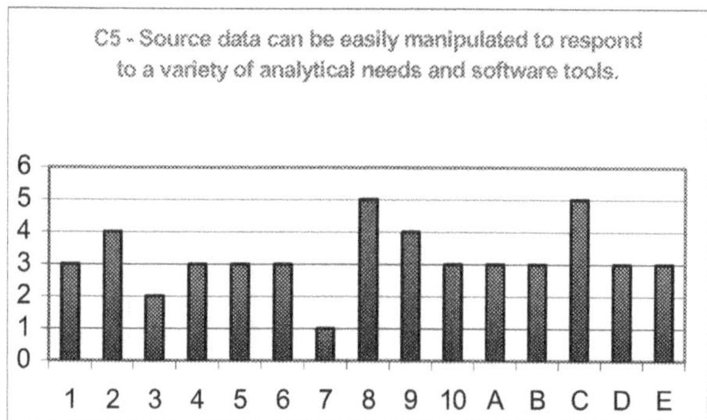
C5 - Source data can be easily manipulated to respond to a variety of analytical needs and software tools.
0 1 2 3 4 5 6
1 2 3 4 5 6 7 8 9 10 A B C D E

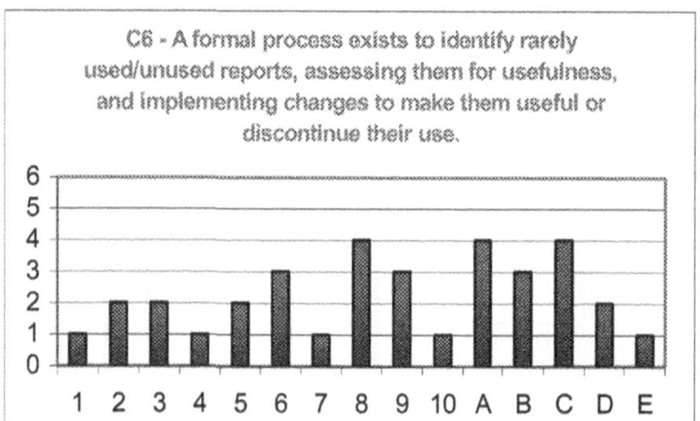
C6 - A formal process exists to identify rarely used/unused reports, assessing them for usefulness, and implementing changes to make them useful or discontinue their use.
0 1 2 3 4 5 6
1 2 3 4 5 6 7 8 9 10 A B C D E

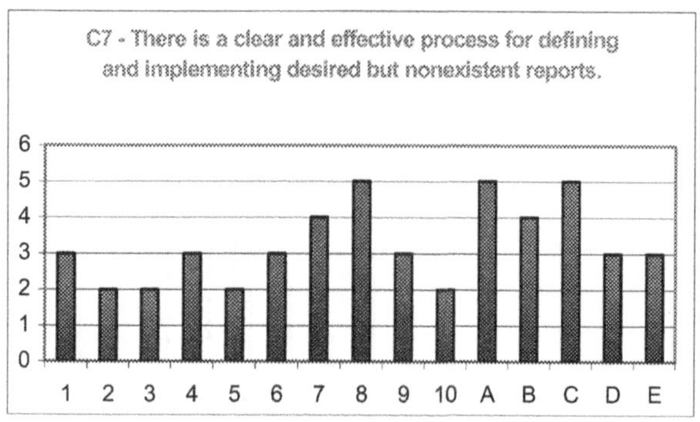
C7 - There is a clear and effective process for defining and implementing desired but nonexistent reports.
0 1 2 3 4 5 6
1 2 3 4 5 6 7 8 9 10 A B C D E

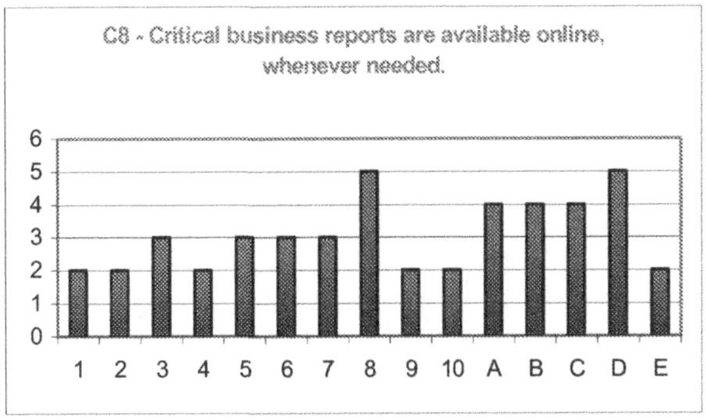


C8 - Critical business reports are available online, whenever needed.

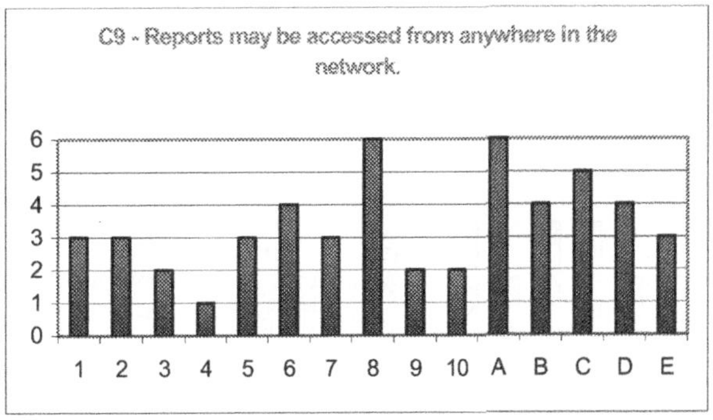


C9 - Reports may be accessed from anywhere in the network.

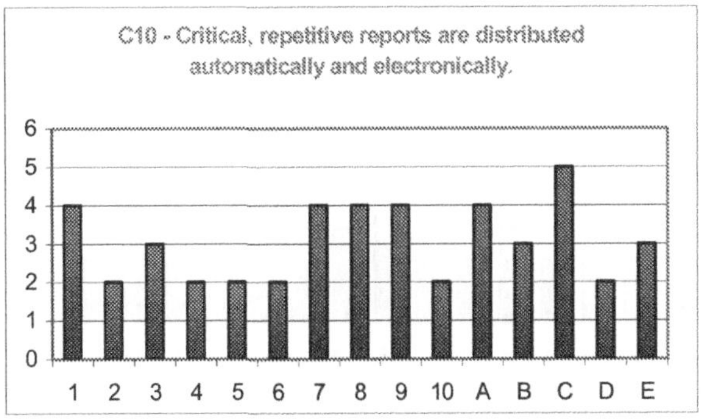


C10 - Critical, repetitive reports are distributed automatically and electronically.

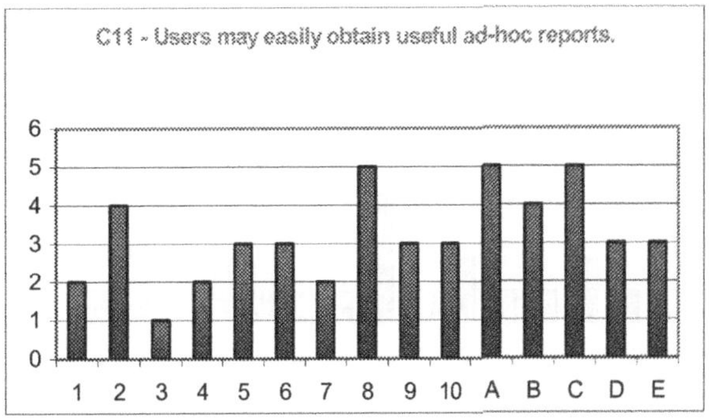


C11 - Users may easily obtain useful ad-hoc reports.

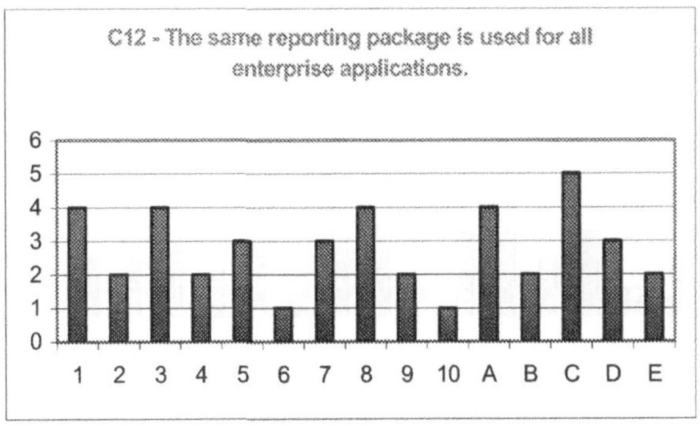


C12 - The same reporting package is used for all enterprise applications.

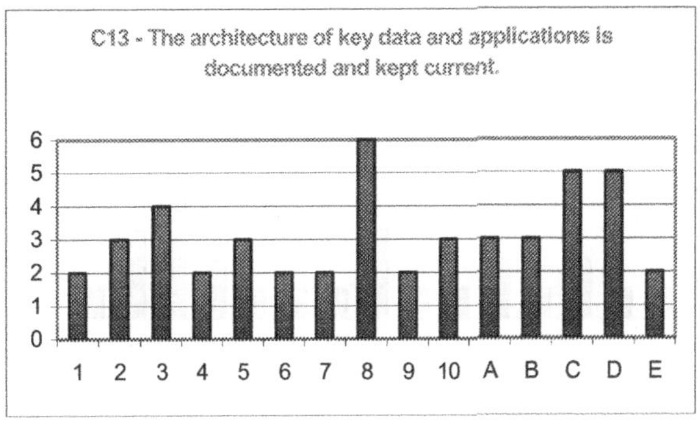


C13 - The architecture of key data and applications is documented and kept current.

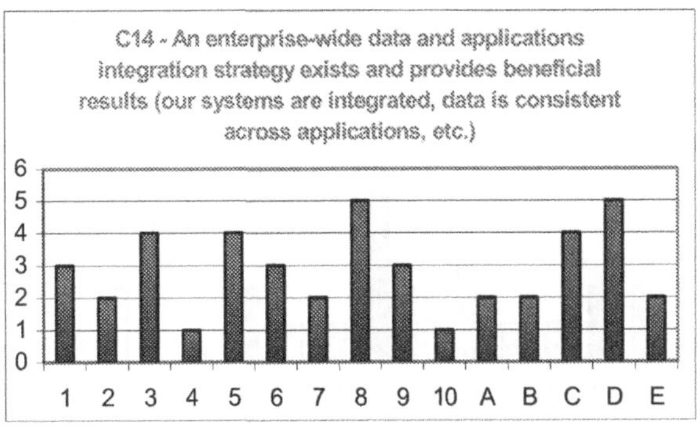


C14 - An enterprise-wide data and applications integration strategy exists and provides beneficial results (our systems are integrated, data is consistent across applications, etc.)

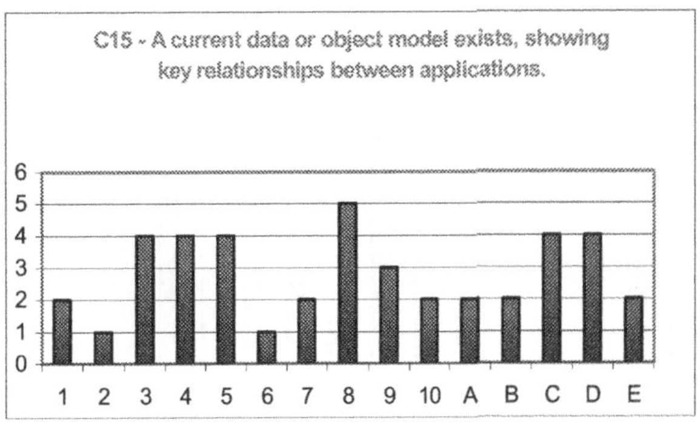


C15 - A current data or object model exists, showing key relationships between applications.

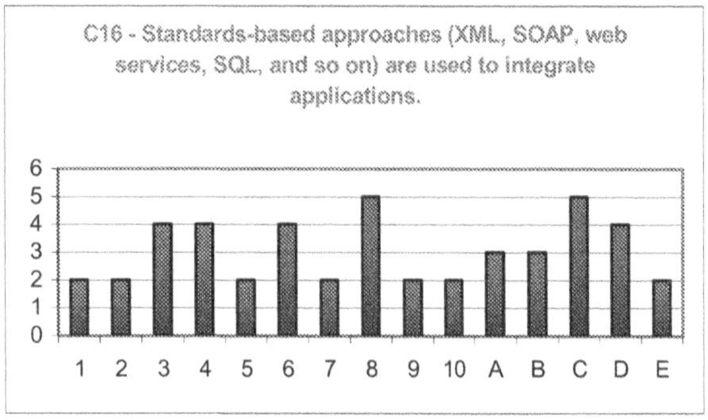
C16 - Standards-based approaches (XML, SOAP, web services, SQL, and so on) are used to integrate applications.
0 1 2 3 4 5 6
1 2 3 4 5 6 7 8 9 10 A B C D E

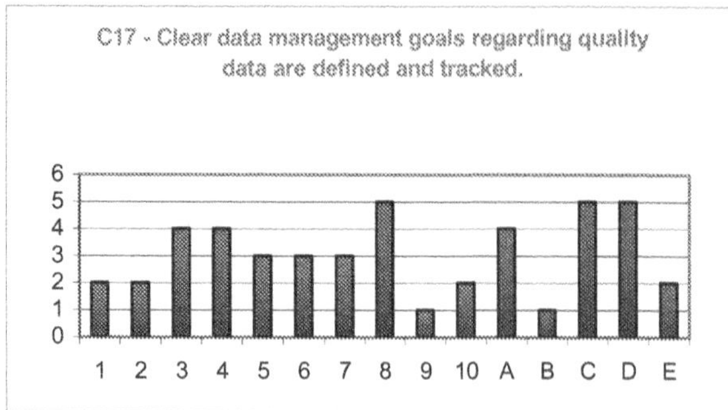
C17 - Clear data management goals regarding quality data are defined and tracked.
0 1 2 3 4 5 6
1 2 3 4 5 6 7 8 9 10 A B C D E

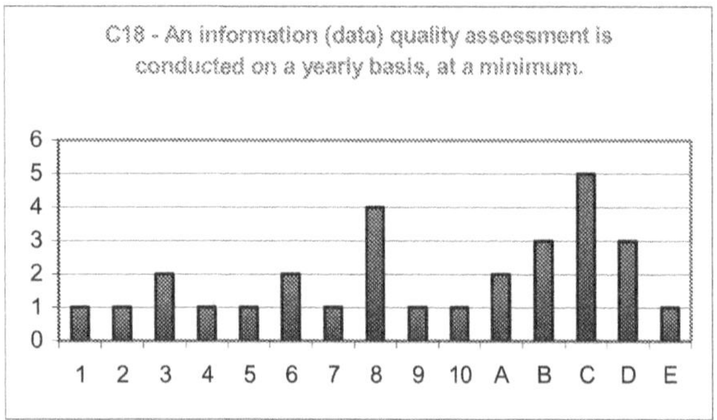
C18 - An information (data) quality assessment is conducted on a yearly basis, at a minimum.
0 1 2 3 4 5 6
1 2 3 4 5 6 7 8 9 10 A B C D E

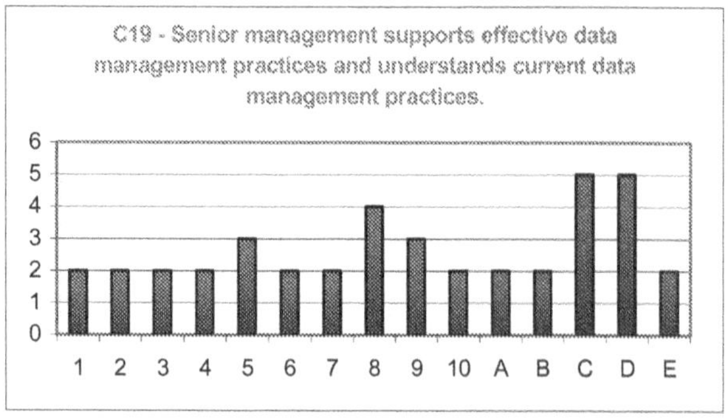
C19 - Senior management supports effective data management practices and understands current data management practices.
0 1 2 3 4 5 6
1 2 3 4 5 6 7 8 9 10 A B C D E

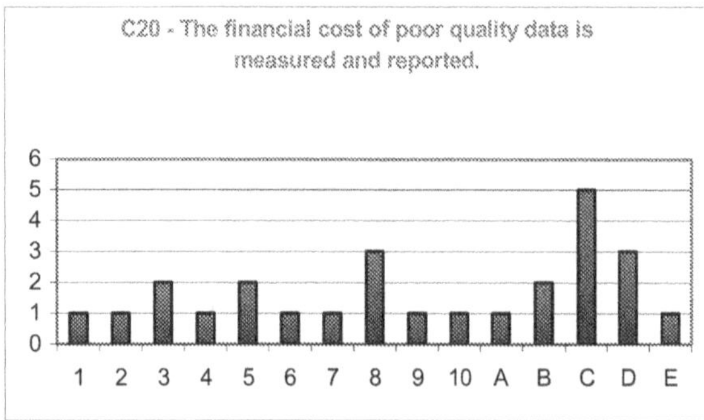
C20 - The financial cost of poor quality data is measured and reported.
0 1 2 3 4 5 6
1 2 3 4 5 6 7 8 9 10 A B C D E

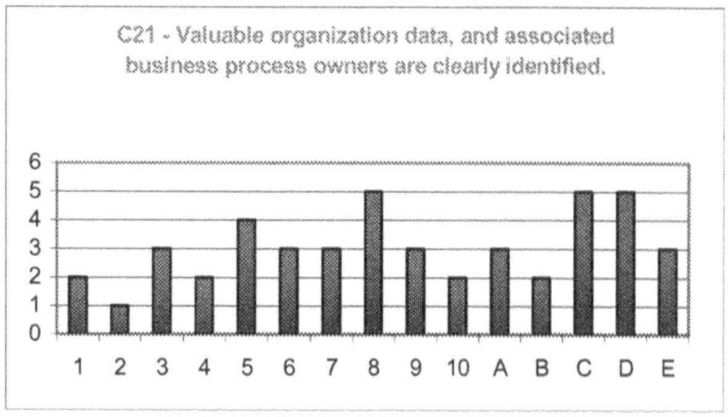
C21 - Valuable organization data, and associated business process owners are clearly identified.
0 1 2 3 4 5 6
1 2 3 4 5 6 7 8 9 10 A B C D E

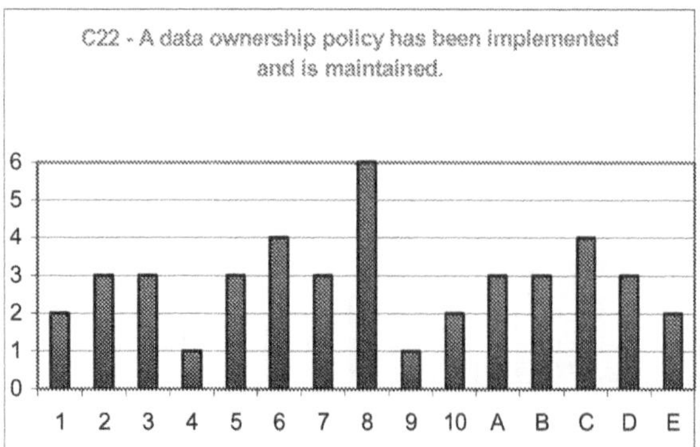
C22 - A data ownership policy has been implemented and is maintained.
0 1 2 3 4 5 6
1 2 3 4 5 6 7 8 9 10 A B C D E

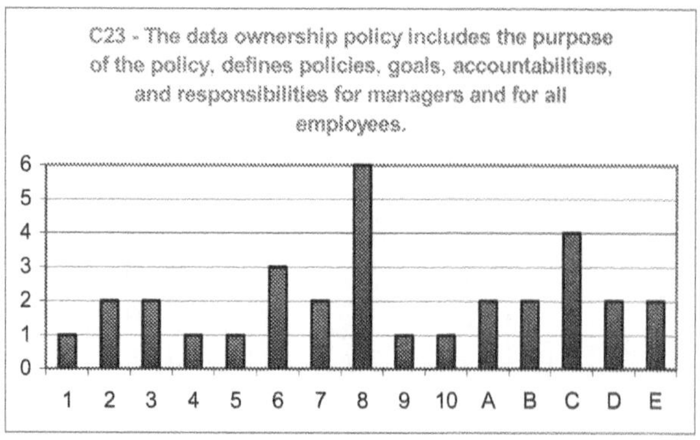
C23 - The data ownership policy includes the purpose of the policy, defines policies, goals, accountabilities, and responsibilities for managers and for all employees.
0 1 2 3 4 5 6
1 2 3 4 5 6 7 8 9 10 A B C D E

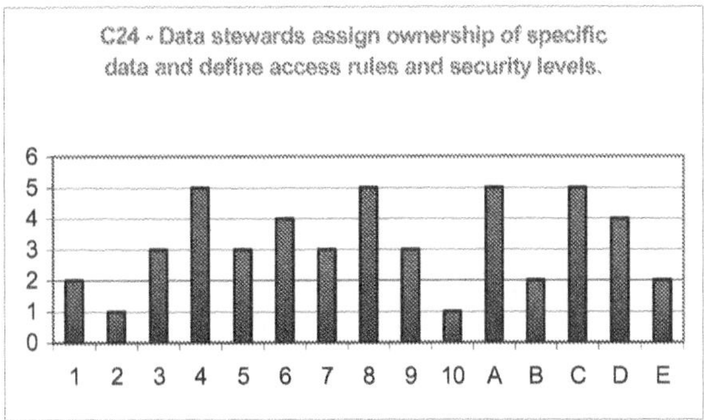
C24 - Data stewards assign ownership of specific data and define access rules and security levels.
0 1 2 3 4 5 6
1 2 3 4 5 6 7 8 9 10 A B C D E

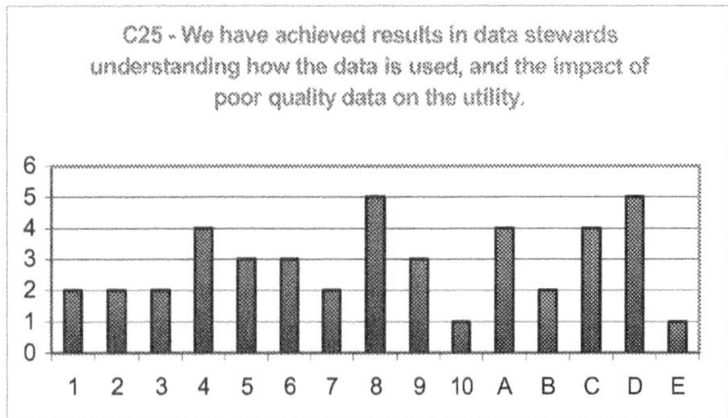
C25 - We have achieved results in data stewards understanding how the data is used, and the impact of poor quality data on the utility.
0 1 2 3 4 5 6
1 2 3 4 5 6 7 8 9 10 A B C D E

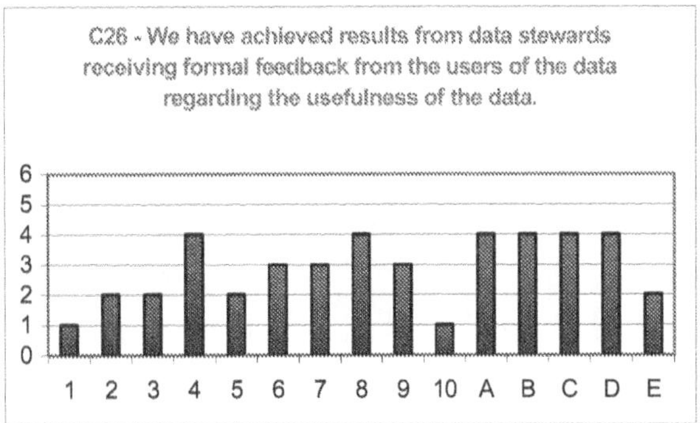
C26 - We have achieved results from data stewards receiving formal feedback from the users of the data regarding the usefulness of the data.
0 1 2 3 4 5 6
1 2 3 4 5 6 7 8 9 10 A B C D E

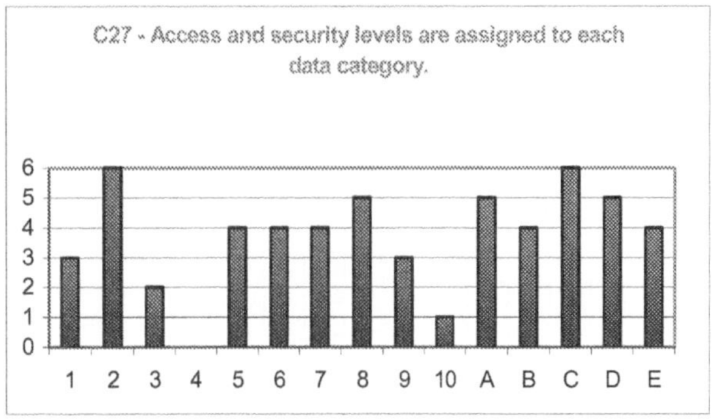
C27 - Access and security levels are assigned to each data category.
0 1 2 3 4 5 6
1 2 3 4 5 6 7 8 9 10 A B C D E

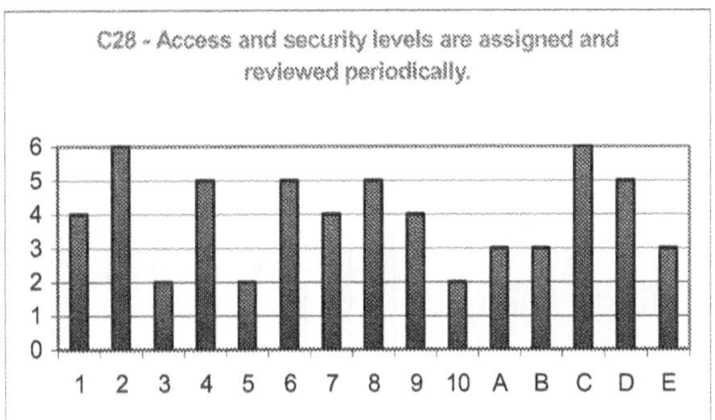
C28 - Access and security levels are assigned and reviewed periodically.
0 1 2 3 4 5 6
1 2 3 4 5 6 7 8 9 10 A B C D E

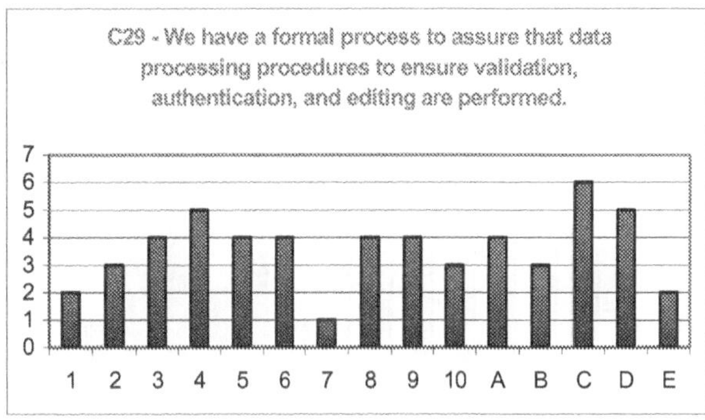
C29 - We have a formal process to assure that data processing procedures to ensure validation, authentication, and editing are performed.
0 1 2 3 4 5 6 7
1 2 3 4 5 6 7 8 9 10 A B C D E

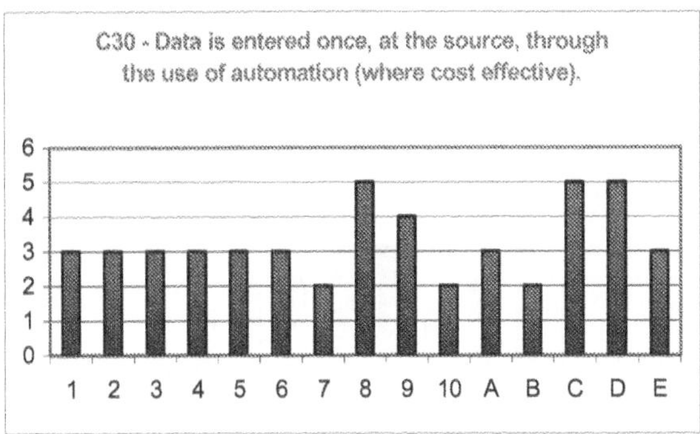
C30 - Data is entered once, at the source, through the use of automation (where cost effective).
0 1 2 3 4 5 6
1 2 3 4 5 6 7 8 9 10 A B C D E

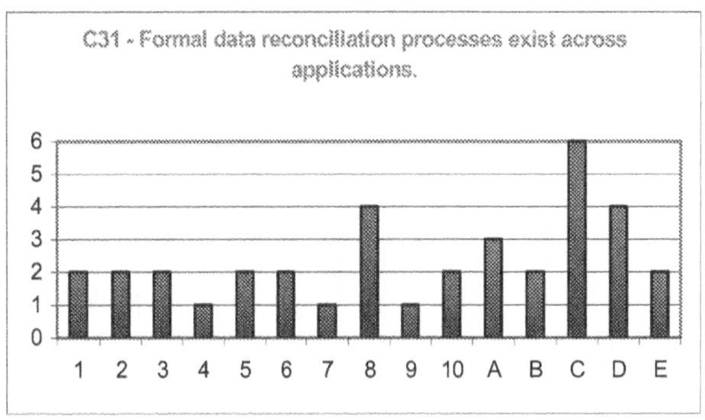
C31 - Formal data reconciliation processes exist across applications.
0 1 2 3 4 5 6
1 2 3 4 5 6 7 8 9 10 A B C D E

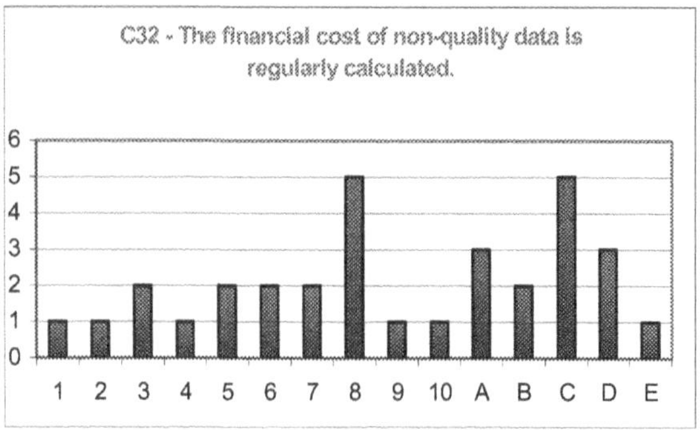
C32 - The financial cost of non-quality data is regularly calculated.
0 1 2 3 4 5 6
1 2 3 4 5 6 7 8 9 10 A B C D E

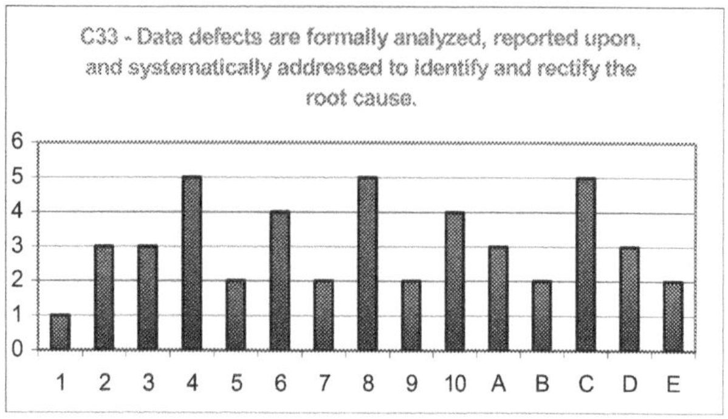
C33 - Data defects are formally analyzed, reported upon, and systematically addressed to identify and rectify the root cause.
0 1 2 3 4 5 6
1 2 3 4 5 6 7 8 9 10 A B C D E

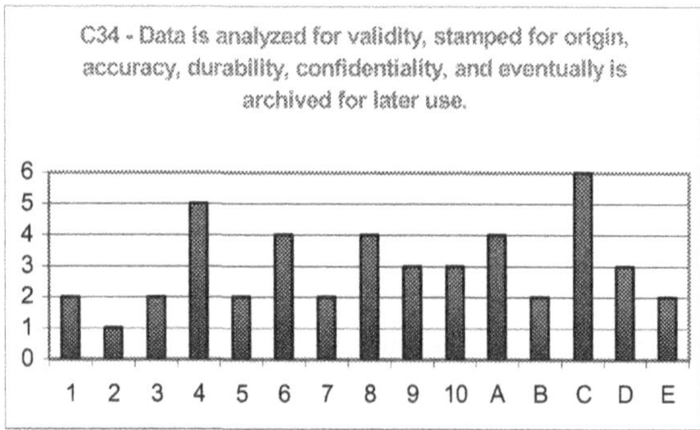
C34 - Data is analyzed for validity, stamped for origin, accuracy, durability, confidentiality, and eventually is archived for later use.
0 1 2 3 4 5 6
1 2 3 4 5 6 7 8 9 10 A B C D E

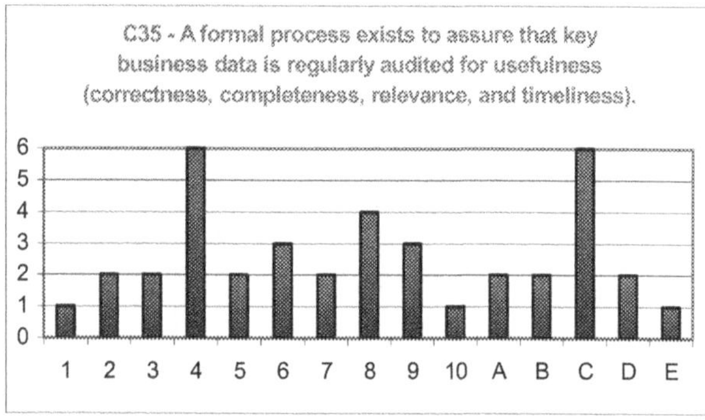
C35 - A formal process exists to assure that key business data is regularly audited for usefulness (correctness, completeness, relevance, and timeliness).
0 1 2 3 4 5 6
1 2 3 4 5 6 7 8 9 10 A B C D E

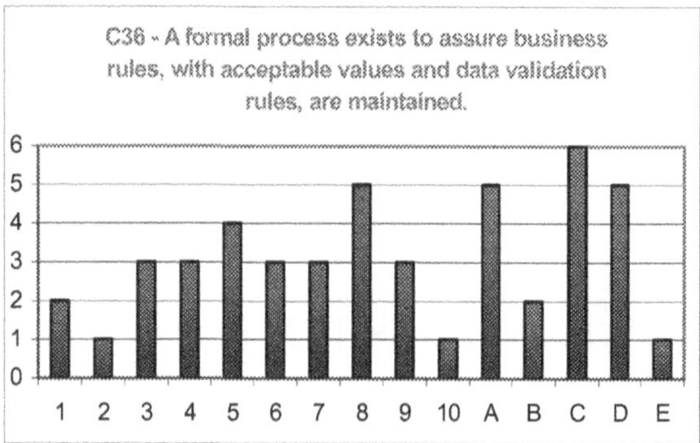
C36 - A formal process exists to assure business rules, with acceptable values and data validation rules, are maintained.
0 1 2 3 4 5 6
1 2 3 4 5 6 7 8 9 10 A B C D E

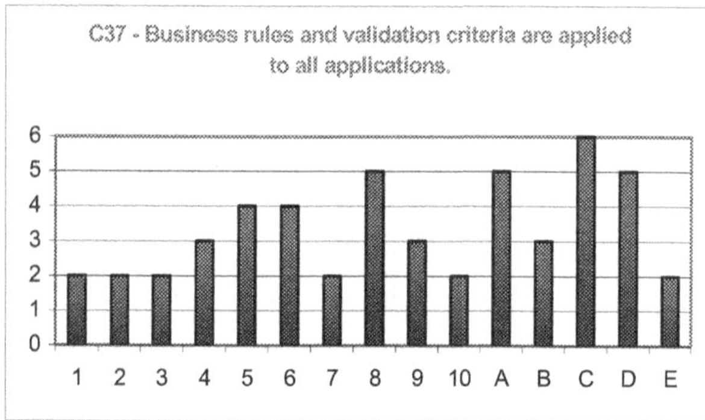
C37 - Business rules and validation criteria are applied to all applications.
0 1 2 3 4 5 6
1 2 3 4 5 6 7 8 9 10 A B C D E

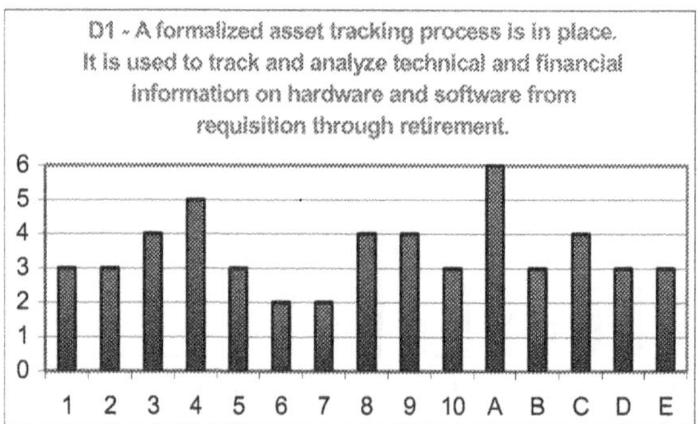
D1 - A formalized asset tracking process is in place. It is used to track and analyze technical and financial information on hardware and software from requisition through retirement.
0 1 2 3 4 5 6
1 2 3 4 5 6 7 8 9 10 A B C D E

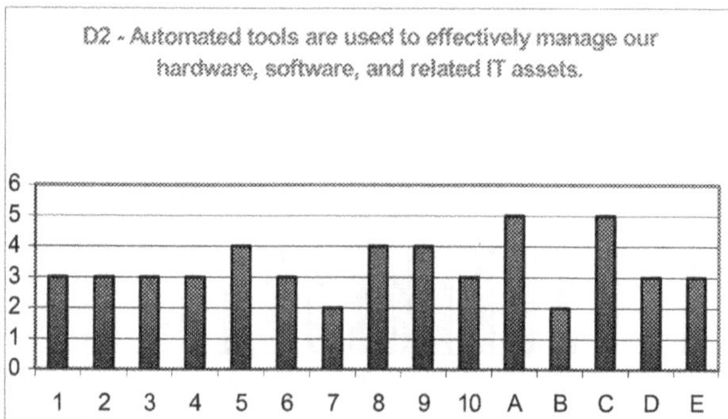
D2 - Automated tools are used to effectively manage our hardware, software, and related IT assets.
0 1 2 3 4 5 6
1 2 3 4 5 6 7 8 9 10 A B C D E

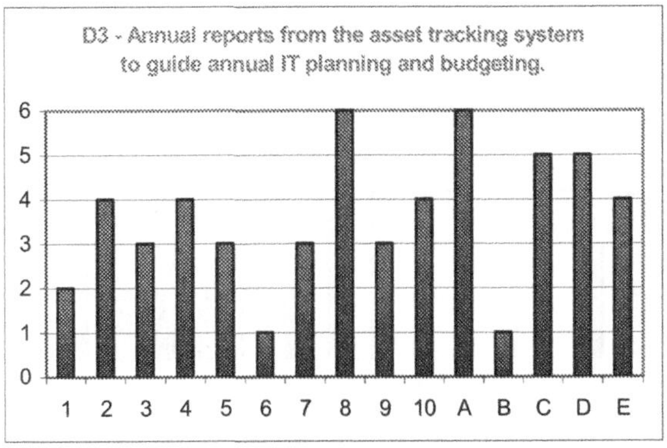
D3 - Annual reports from the asset tracking system
to guide annual IT planning and budgeting.
0 1 2 3 4 5 6
1 2 3 4 5 6 7 8 9 10 A B C D E

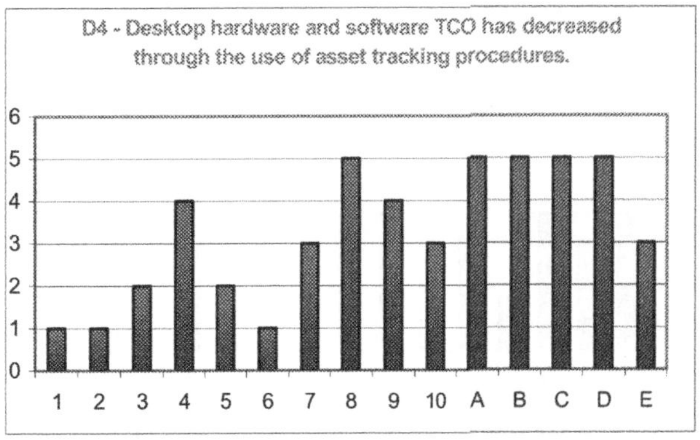
D4 - Desktop hardware and software TCO has decreased
through the use of asset tracking procedures.
0 1 2 3 4 5 6
1 2 3 4 5 6 7 8 9 10 A B C D E

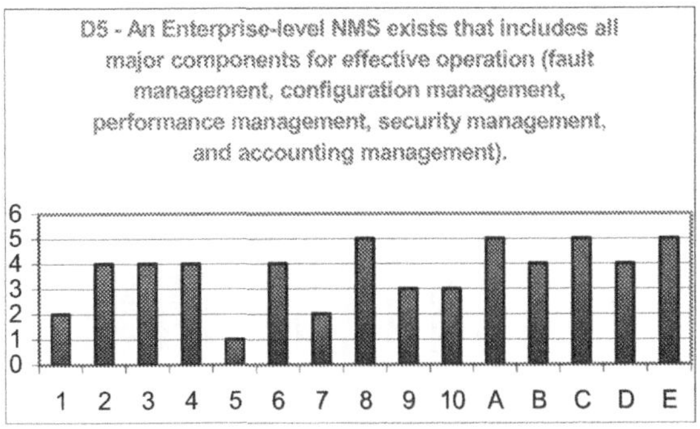
D5 - An Enterprise-level NMS exists that includes all
major components for effective operation (fault
management, configuration management,
performance management, security management,
and accounting management).
0 1 2 3 4 5 6
1 2 3 4 5 6 7 8 9 10 A B C D E

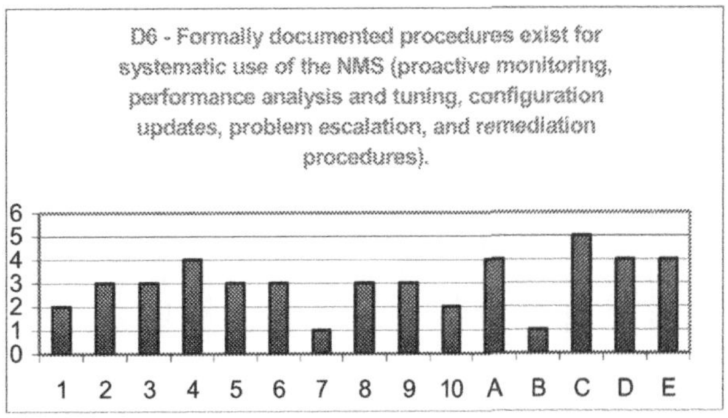
D6 - Formally documented procedures exist for
systematic use of the NMS (proactive monitoring,
performance analysis and tuning, configuration
updates, problem escalation, and remediation
procedures).
0 1 2 3 4 5 6
1 2 3 4 5 6 7 8 9 10 A B C D E

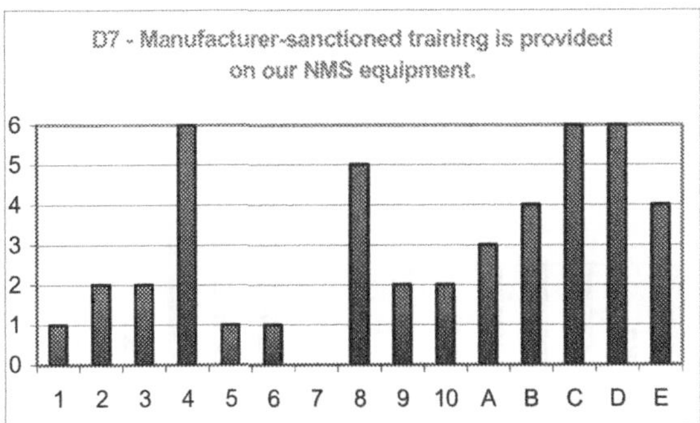
D7 - Manufacturer-sanctioned training is provided
on our NMS equipment.
0 1 2 3 4 5 6
1 2 3 4 5 6 7 8 9 10 A B C D E

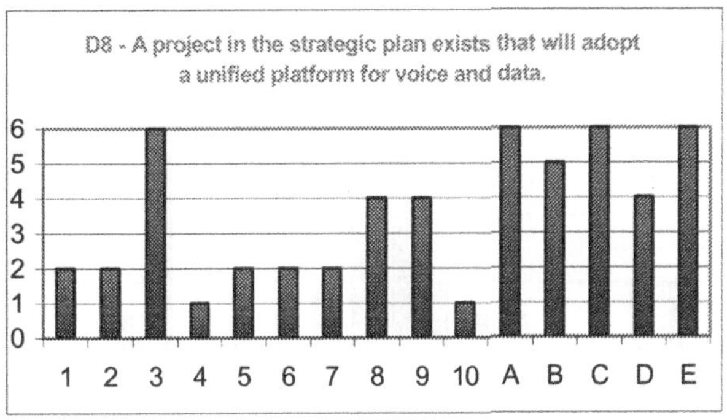
D8 - A project in the strategic plan exists that will adopt
a unified platform for voice and data.
0 1 2 3 4 5 6
1 2 3 4 5 6 7 8 9 10 A B C D E

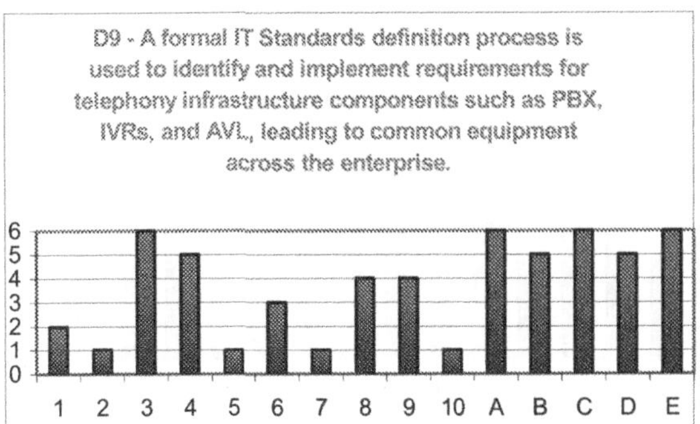
D9 - A formal IT Standards definition process is
used to identify and implement requirements for
telephony infrastructure components such as PBX,
IVRs, and AVL, leading to common equipment
across the enterprise.
0 1 2 3 4 5 6
1 2 3 4 5 6 7 8 9 10 A B C D E

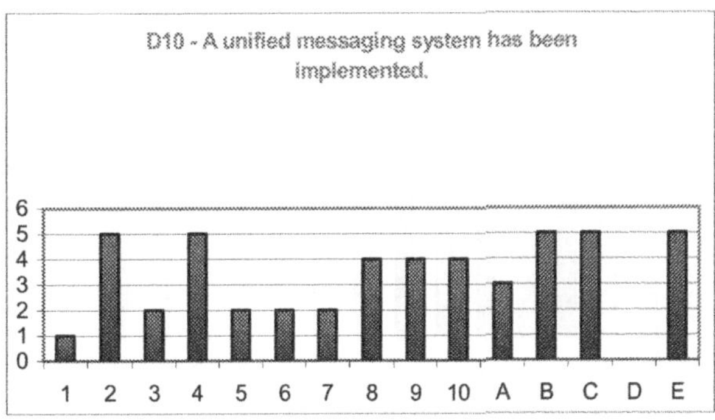
D10 - A unified messaging system has been
implemented.
0 1 2 3 4 5 6
1 2 3 4 5 6 7 8 9 10 A B C D E

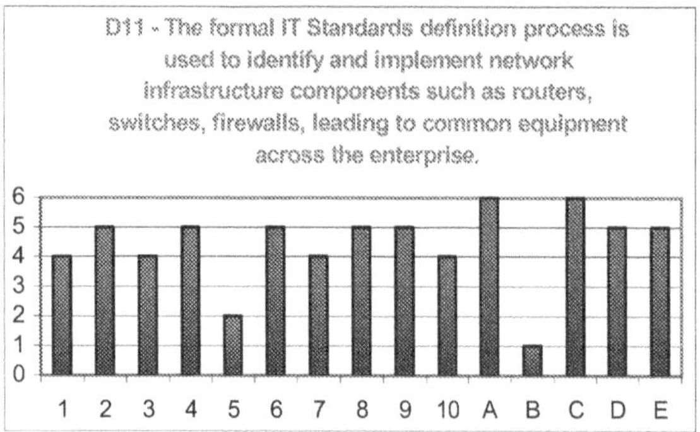
D11 - The formal IT Standards definition process is used to identify and implement network infrastructure components such as routers, switches, firewalls, leading to common equipment across the enterprise.
0 1 2 3 4 5 6
1 2 3 4 5 6 7 8 9 10 A B C D E

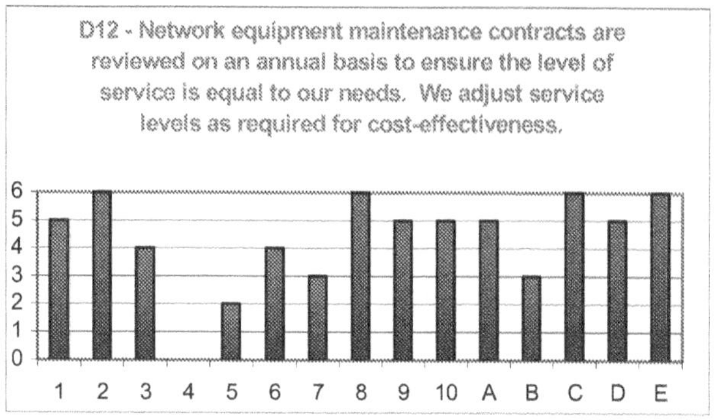
D12 - Network equipment maintenance contracts are reviewed on an annual basis to ensure the level of service is equal to our needs. We adjust service levels as required for cost-effectiveness.
0 1 2 3 4 5 6
1 2 3 4 5 6 7 8 9 10 A B C D E

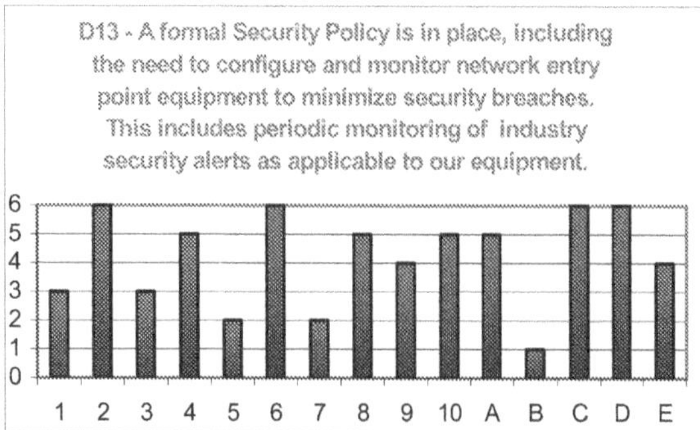
D13 - A formal Security Policy is in place, including the need to configure and monitor network entry point equipment to minimize security breaches. This includes periodic monitoring of industry security alerts as applicable to our equipment.
0 1 2 3 4 5 6
1 2 3 4 5 6 7 8 9 10 A B C D E

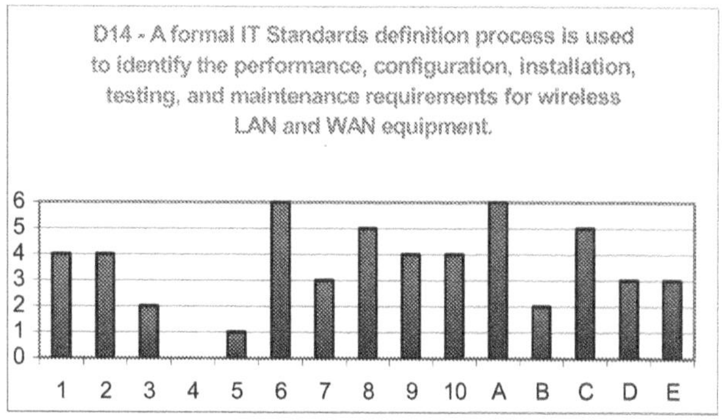
D14 - A formal IT Standards definition process is used to identify the performance, configuration, installation, testing, and maintenance requirements for wireless LAN and WAN equipment.
0 1 2 3 4 5 6
1 2 3 4 5 6 7 8 9 10 A B C D E

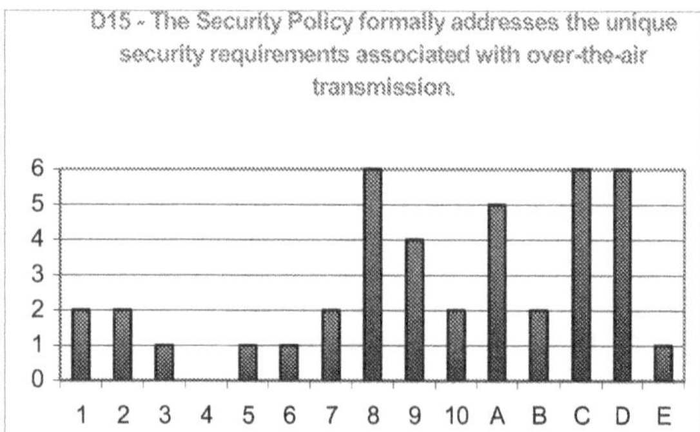
D15 - The Security Policy formally addresses the unique security requirements associated with over-the-air transmission.
0 1 2 3 4 5 6
1 2 3 4 5 6 7 8 9 10 A B C D E

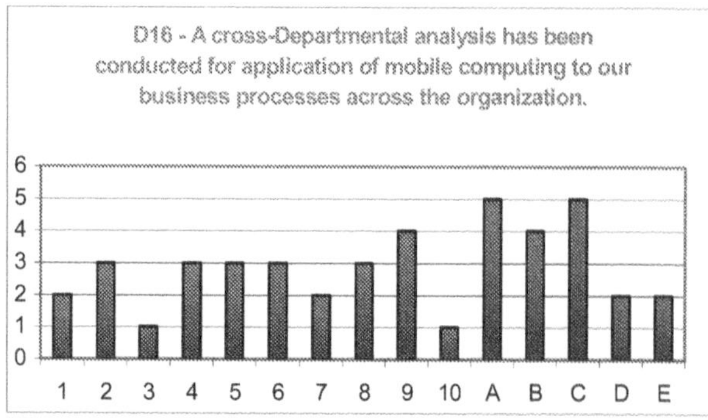
D16 - A cross-Departmental analysis has been conducted for application of mobile computing to our business processes across the organization.
0 1 2 3 4 5 6
1 2 3 4 5 6 7 8 9 10 A B C D E

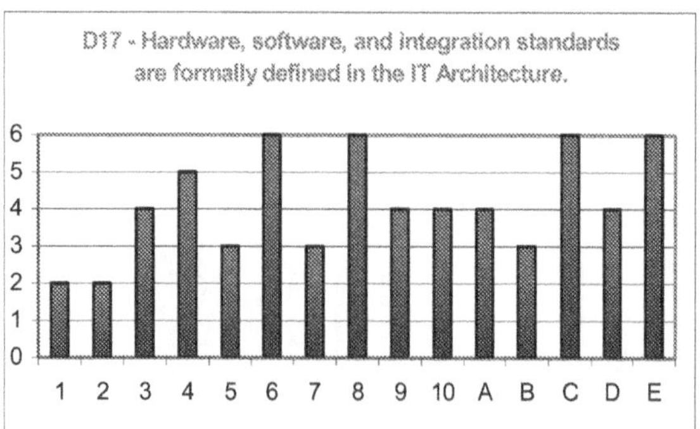
D17 - Hardware, software, and integration standards are formally defined in the IT Architecture.
0 1 2 3 4 5 6
1 2 3 4 5 6 7 8 9 10 A B C D E

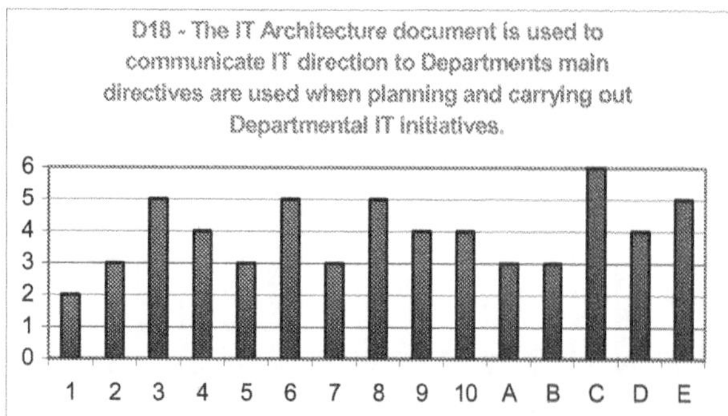
D18 - The IT Architecture document is used to communicate IT direction to Departments main directives are used when planning and carrying out Departmental IT initiatives.
0 1 2 3 4 5 6
1 2 3 4 5 6 7 8 9 10 A B C D E

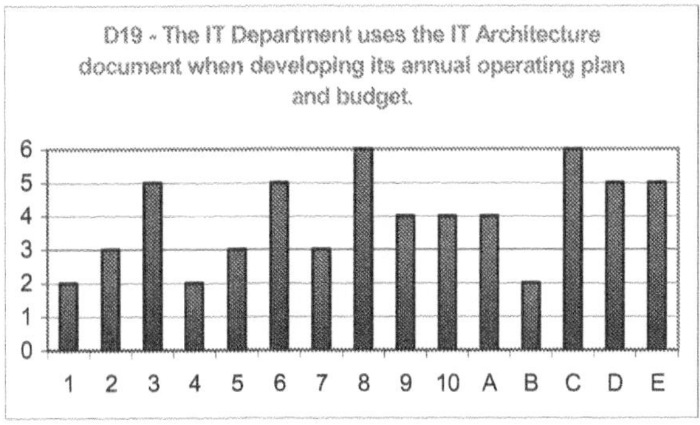
D19 - The IT Department uses the IT Architecture document when developing its annual operating plan and budget.
0 1 2 3 4 5 6
1 2 3 4 5 6 7 8 9 10 A B C D E

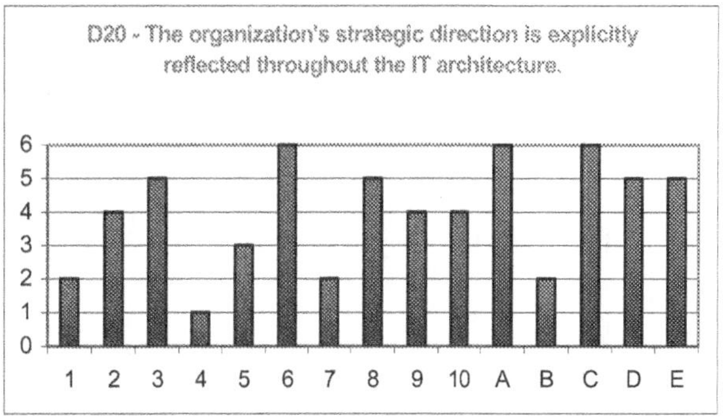
D20 - The organization's strategic direction is explicitly reflected throughout the IT architecture.
0 1 2 3 4 5 6
1 2 3 4 5 6 7 8 9 10 A B C D E

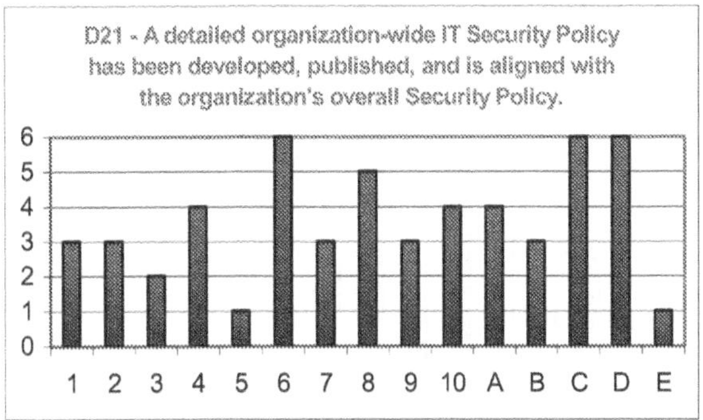
D21 - A detailed organization-wide IT Security Policy has been developed, published, and is aligned with the organization's overall Security Policy.
0 1 2 3 4 5 6
1 2 3 4 5 6 7 8 9 10 A B C D E

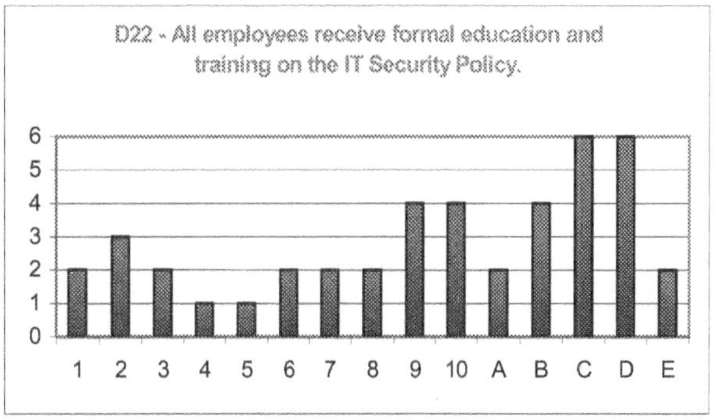
D22 - All employees receive formal education and training on the IT Security Policy.
0 1 2 3 4 5 6
1 2 3 4 5 6 7 8 9 10 A B C D E

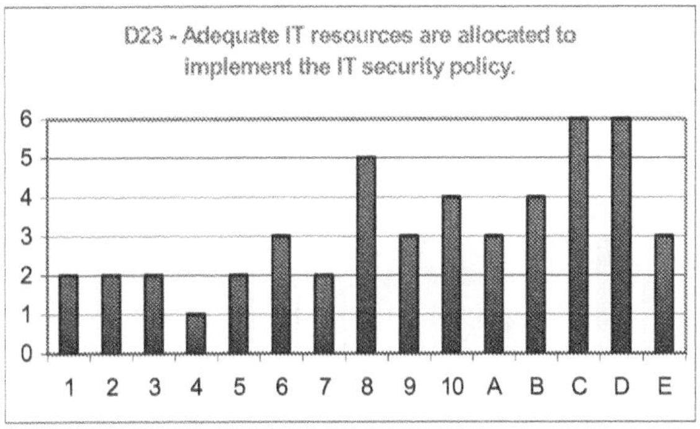
D23 - Adequate IT resources are allocated to implement the IT security policy.
0 1 2 3 4 5 6
1 2 3 4 5 6 7 8 9 10 A B C D E

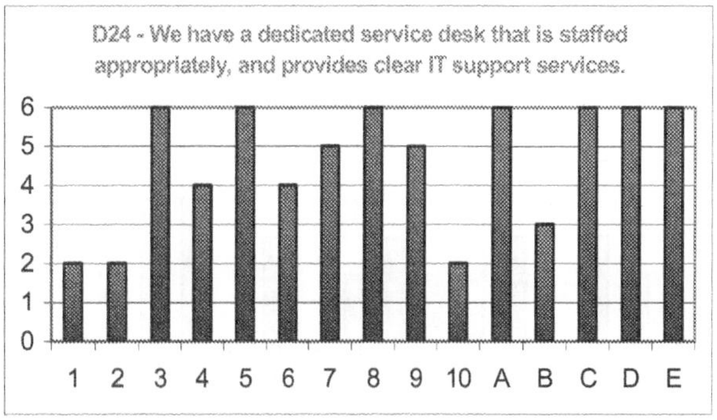
D24 - We have a dedicated service desk that is staffed appropriately, and provides clear IT support services.
0 1 2 3 4 5 6
1 2 3 4 5 6 7 8 9 10 A B C D E

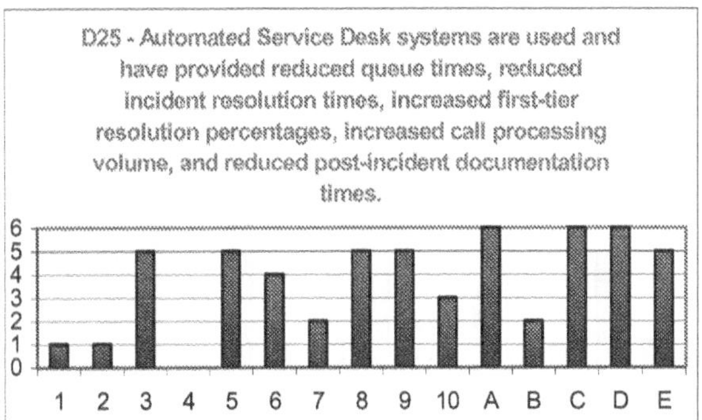
D25 - Automated Service Desk systems are used and have provided reduced queue times, reduced incident resolution times, increased first-tier resolution percentages, increased call processing volume, and reduced post-incident documentation times.
0 1 2 3 4 5 6
1 2 3 4 5 6 7 8 9 10 A B C D E

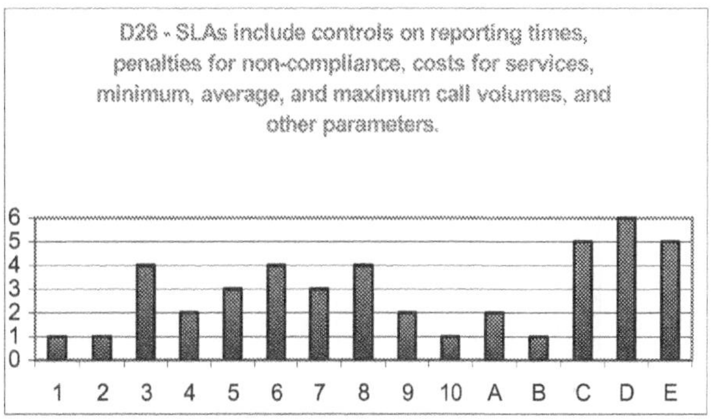
D26 - SLAs include controls on reporting times, penalties for non-compliance, costs for services, minimum, average, and maximum call volumes, and other parameters.
0 1 2 3 4 5 6
1 2 3 4 5 6 7 8 9 10 A B C D E

D27 - Overall Service Desk performance is periodically analyzed so as to improve the levels of service (e.g., offering classes on the most frequently reported problems).

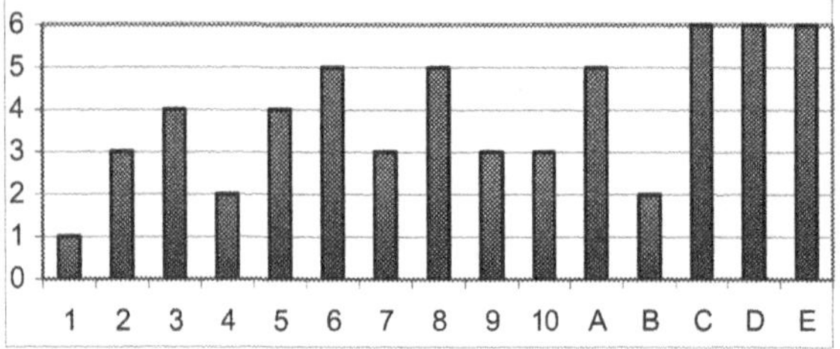


E1 - A formal outsourcing decision-making framework exists to analyze service delivery alternatives.

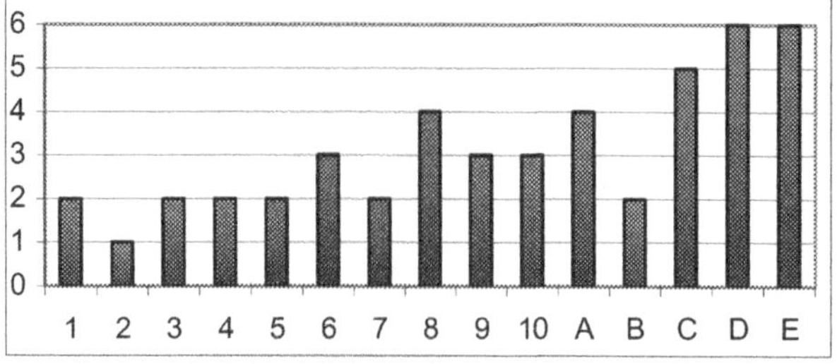


E2 - All third party service providers have well defined goals, objectives, and deliverables.

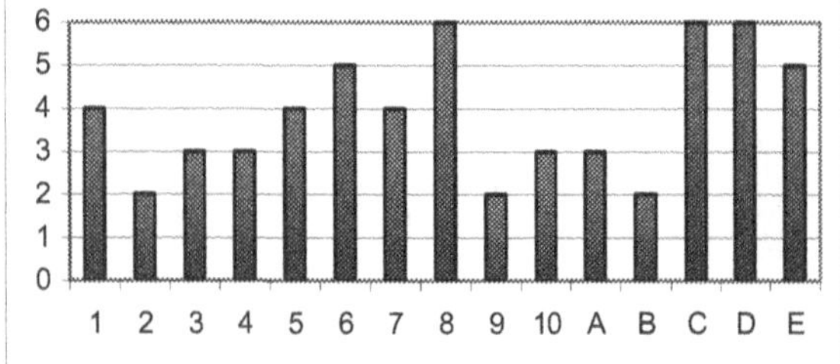


E3 - The performance of third party suppliers is formally reviewed at pre-defined intervals.

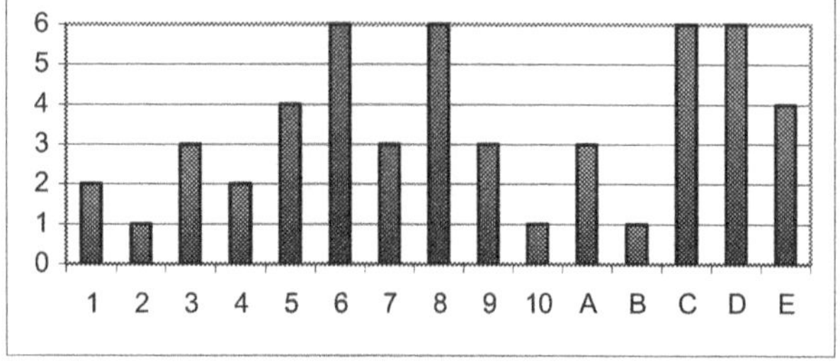


E4 - Measurable cost savings or productivity improvements are achieved by using third party suppliers.

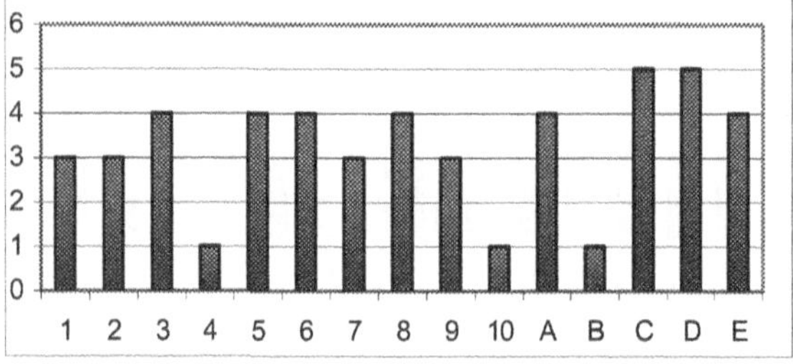


E5 - The use of third party suppliers is periodically reviewed against changes to the IT strategic plan or in market conditions.

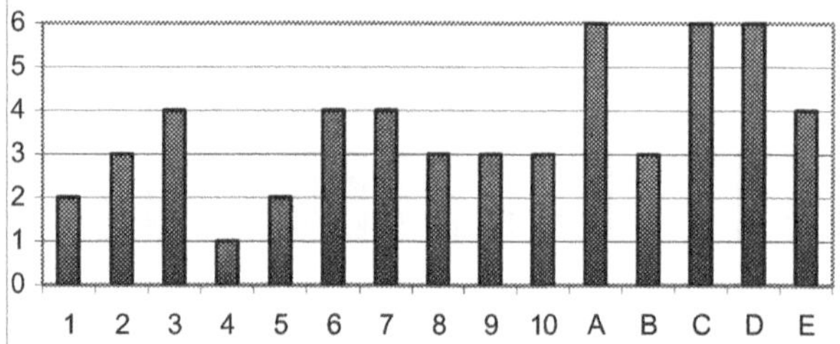


E6 - A formal methodology for defining, negotiating, agreeing, implementing, monitoring and managing the levels of customer service exists.

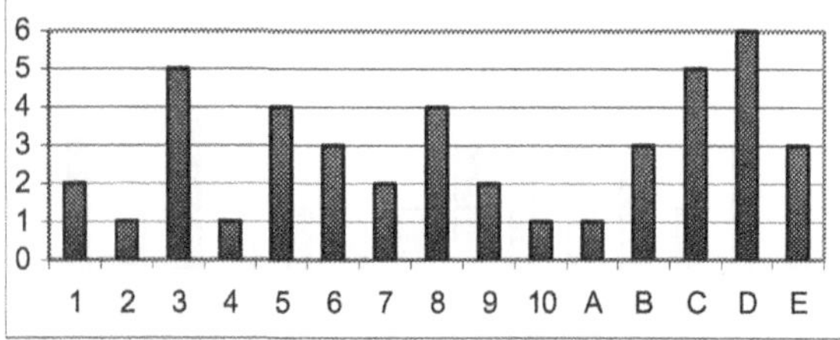


E7 - A formal corrective action process exists for tracking, trending, analyzing and correcting service levels not meeting requirements.

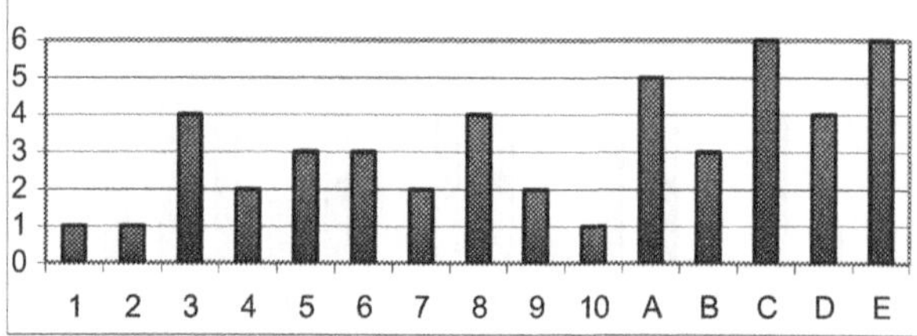

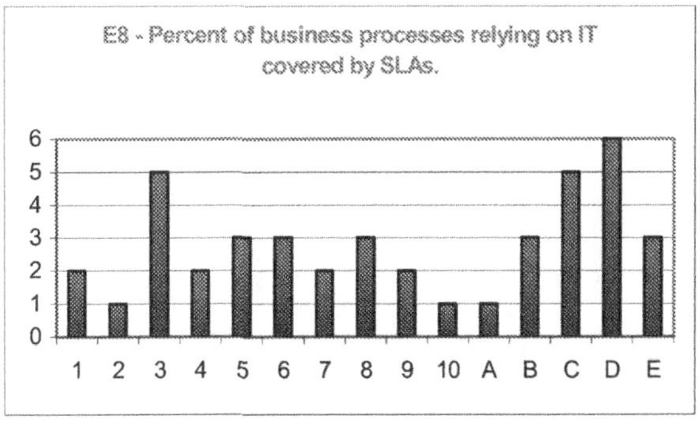
E8 - Percent of business processes relying on IT covered by SLAs.

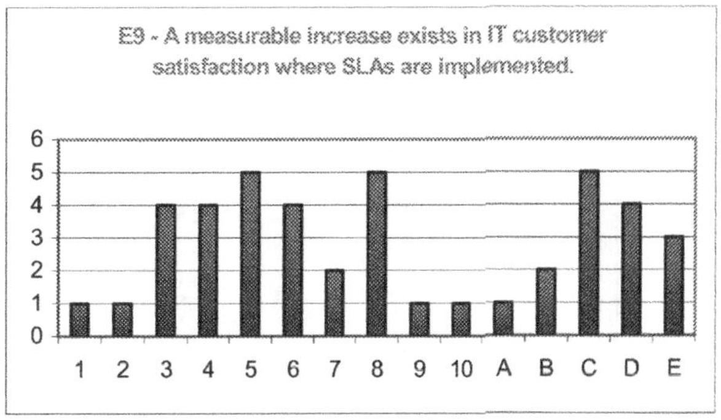
E9 - A measurable increase exists in IT customer satisfaction where SLAs are implemented.

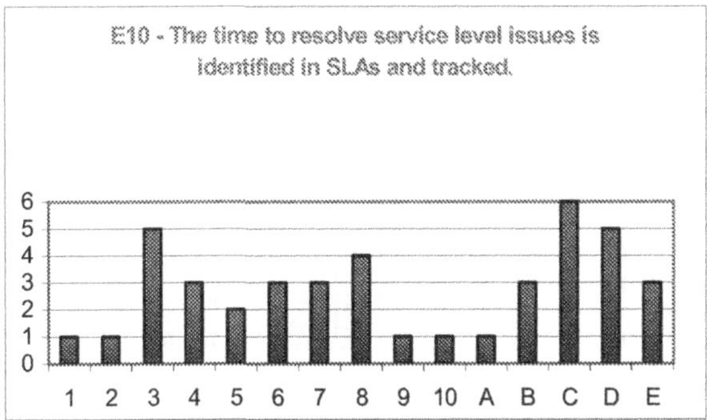
E10 - The time to resolve service level issues is identified in SLAs and tracked.

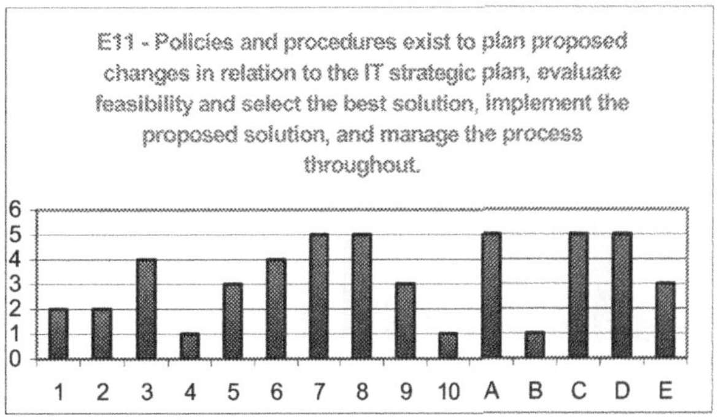
E11 - Policies and procedures exist to plan proposed changes in relation to the IT strategic plan, evaluate feasibility and select the best solution, implement the proposed solution, and manage the process throughout.

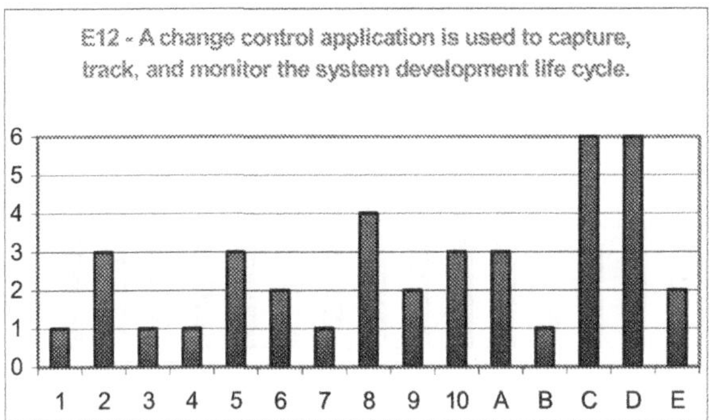
E12 - A change control application is used to capture, track, and monitor the system development life cycle.

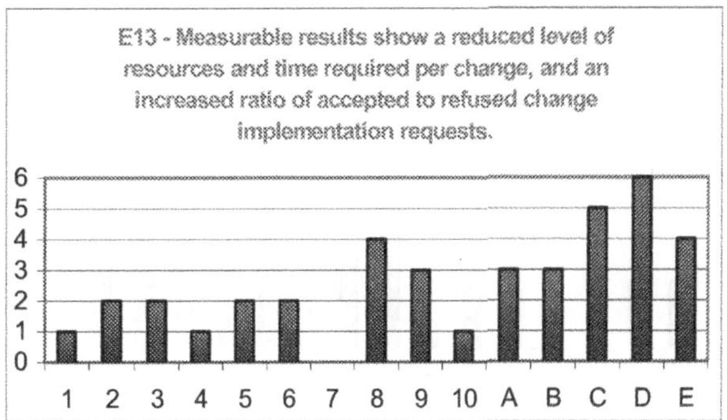
E13 - Measurable results show a reduced level of resources and time required per change, and an increased ratio of accepted to refused change implementation requests.

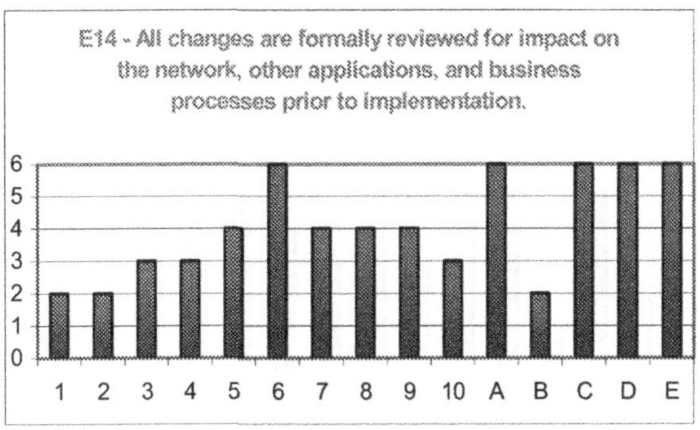
E14 - All changes are formally reviewed for impact on the network, other applications, and business processes prior to implementation.

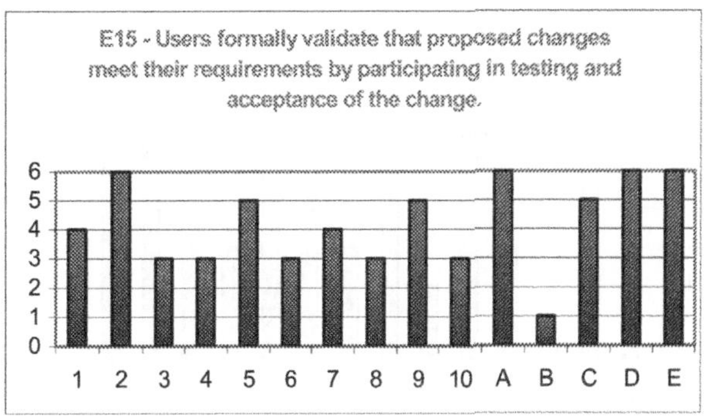
E15 - Users formally validate that proposed changes meet their requirements by participating in testing and acceptance of the change.

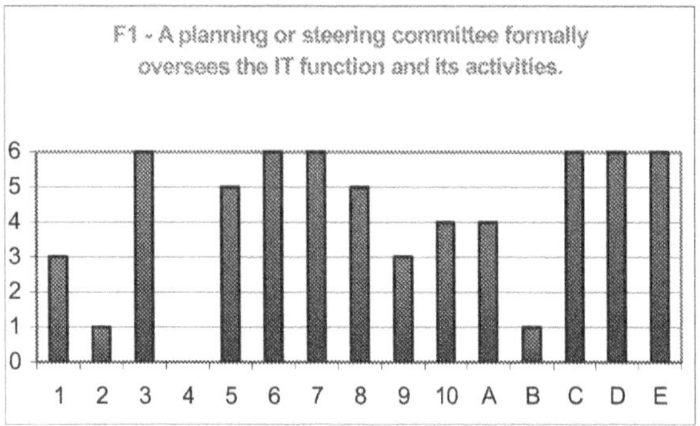
F1 - A planning or steering committee formally oversees the IT function and its activities.
0 1 2 3 4 5 6
1 2 3 4 5 6 7 8 9 10 A B C D E

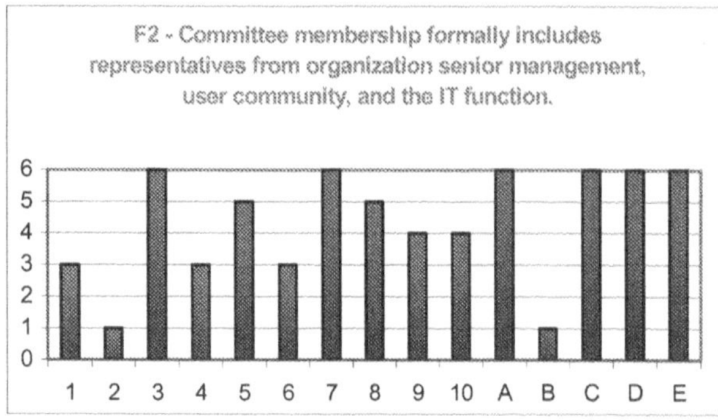
F2 - Committee membership formally includes representatives from organization senior management, user community, and the IT function.
0 1 2 3 4 5 6
1 2 3 4 5 6 7 8 9 10 A B C D E

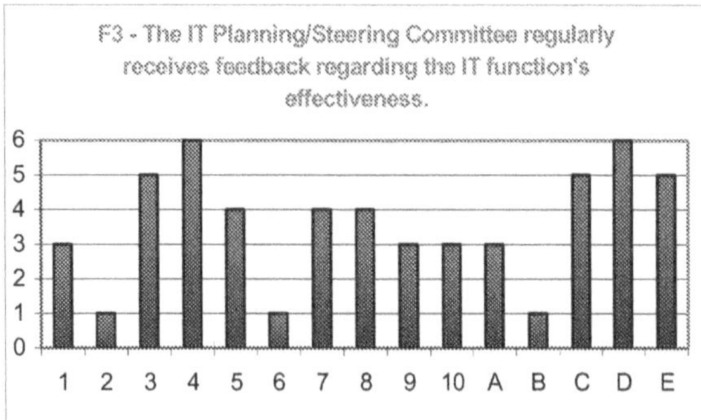
F3 - The IT Planning/Steering Committee regularly receives feedback regarding the IT function's effectiveness.
0 1 2 3 4 5 6
1 2 3 4 5 6 7 8 9 10 A B C D E

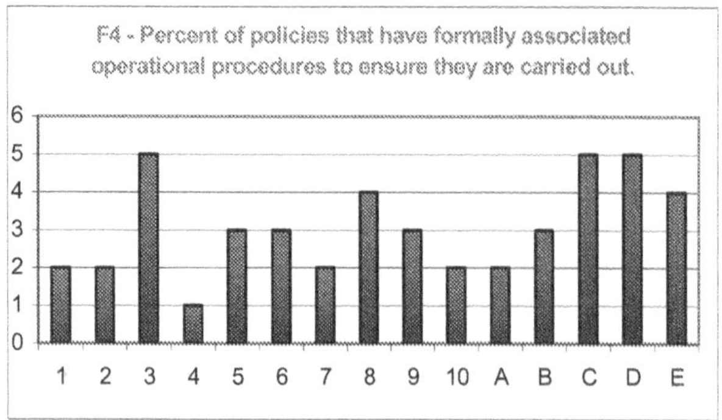
F4 - Percent of policies that have formally associated operational procedures to ensure they are carried out.
0 1 2 3 4 5 6
1 2 3 4 5 6 7 8 9 10 A B C D E

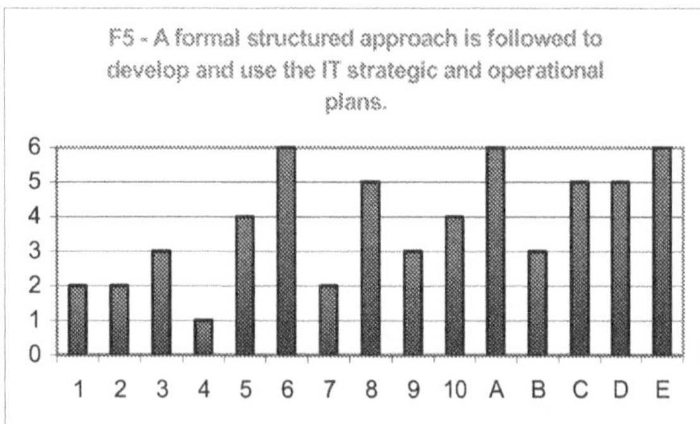
F5 - A formal structured approach is followed to develop and use the IT strategic and operational plans.
0 1 2 3 4 5 6
1 2 3 4 5 6 7 8 9 10 A B C D E

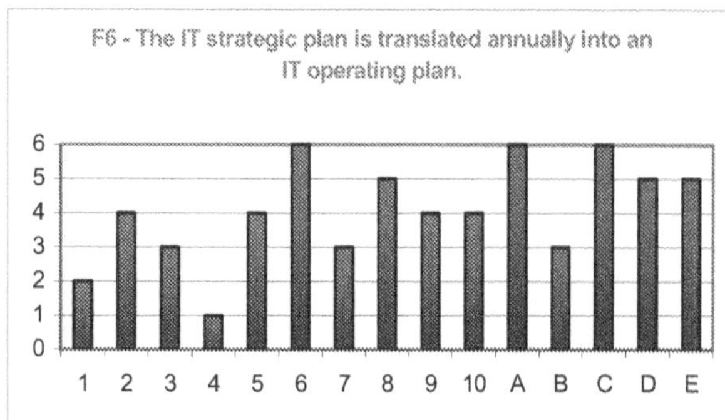
F6 - The IT strategic plan is translated annually into an IT operating plan.
0 1 2 3 4 5 6
1 2 3 4 5 6 7 8 9 10 A B C D E

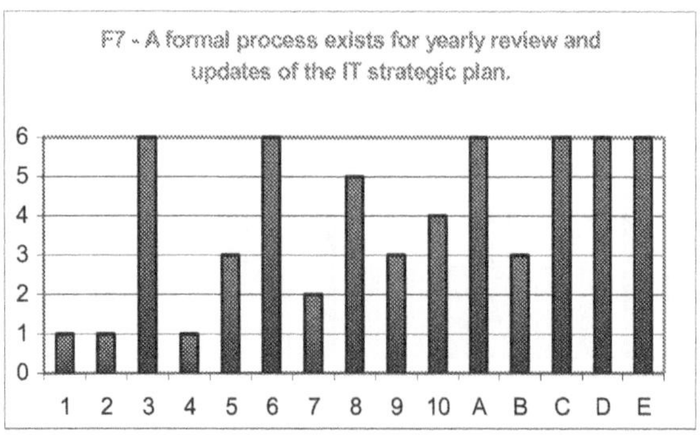
F7 - A formal process exists for yearly review and updates of the IT strategic plan.
0 1 2 3 4 5 6
1 2 3 4 5 6 7 8 9 10 A B C D E

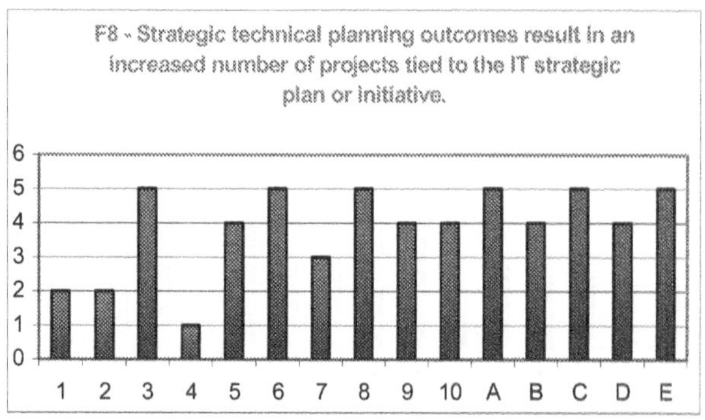
F8 - Strategic technical planning outcomes result in an increased number of projects tied to the IT strategic plan or initiative.
0 1 2 3 4 5 6
1 2 3 4 5 6 7 8 9 10 A B C D E

F9 - IT staff participate in the organization's strategic planning efforts.

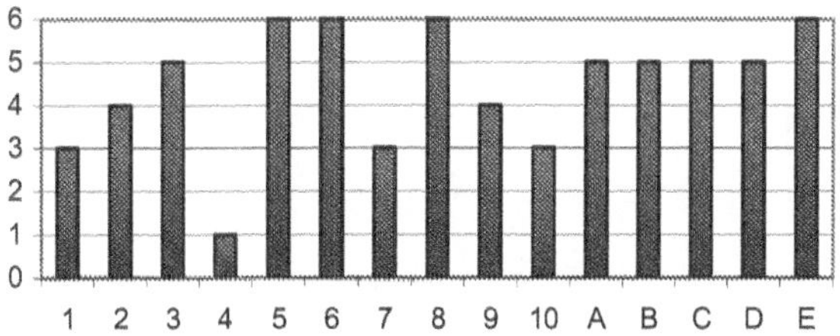


F10 - An annual IT budgeting process is implemented and carried out with significant exceptions to the budget being approved and tracked.

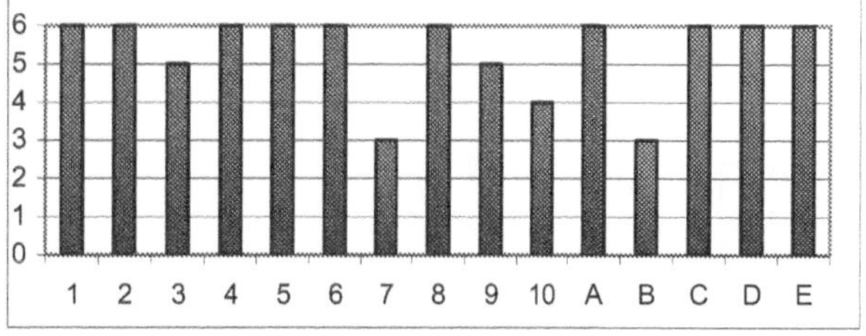


F11 - The annual IT operations budget reflects the organization and IT strategic plans.

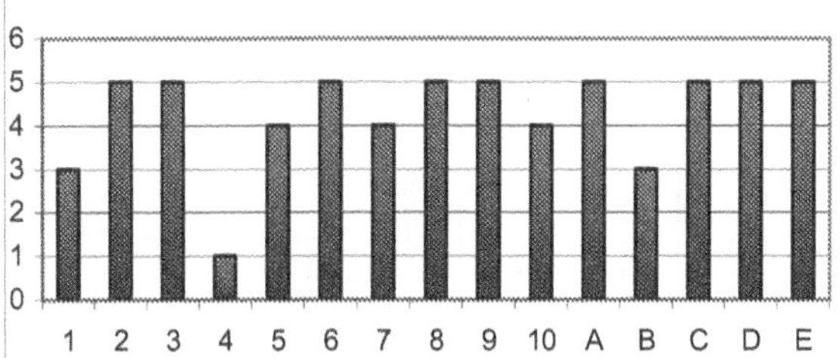


F12 - Variances between strategic plan forecasts and actual costs are analyzed and reported to facilitate decision making.

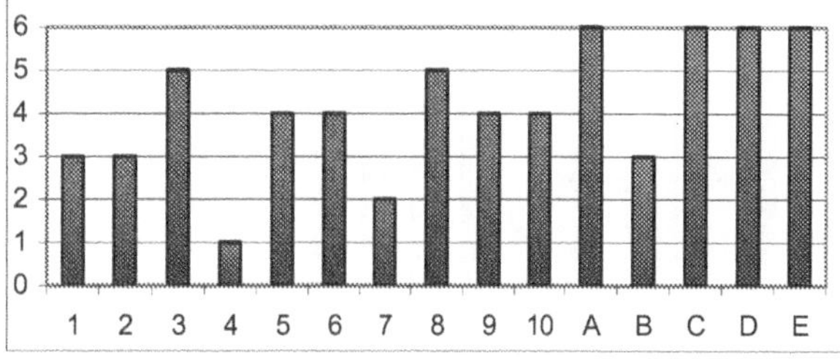


F13 - An automated accounting system exists to support cost monitoring of IT assets products and services.

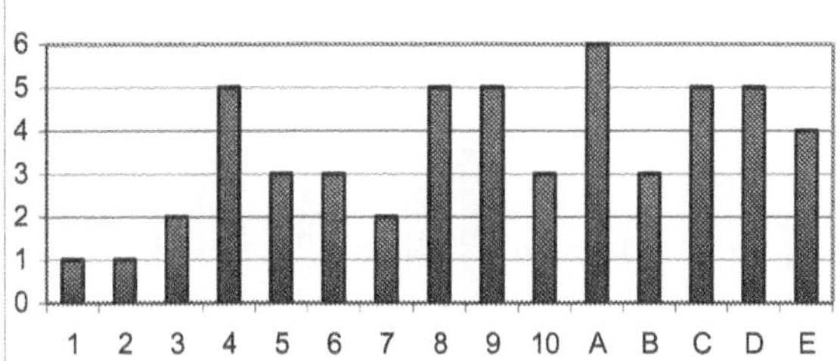


F14 - Whether the system is activity-based or not, costs can be linked to the activities that generate the cost so that proper controls can be established and decisions made.

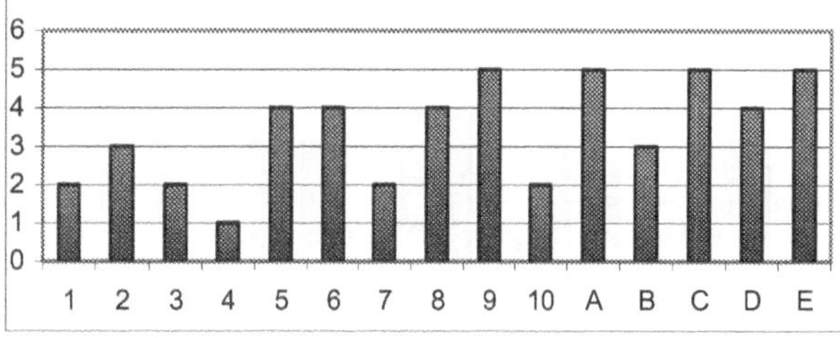


F15 - A formal budget review (planned versus actual) is conducted periodically.

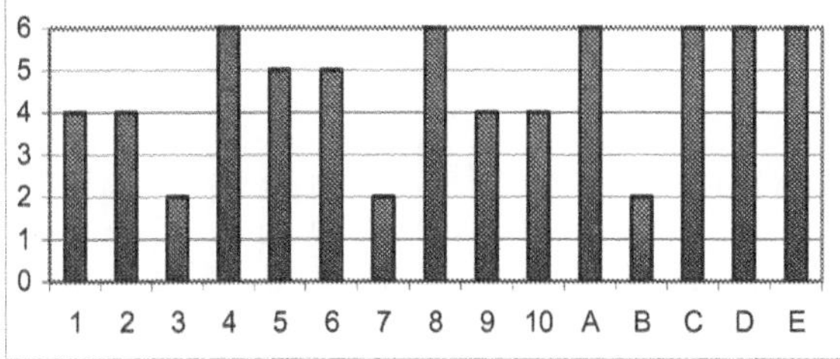


F16 - Actual IT project costs are within 10% of budget estimates.

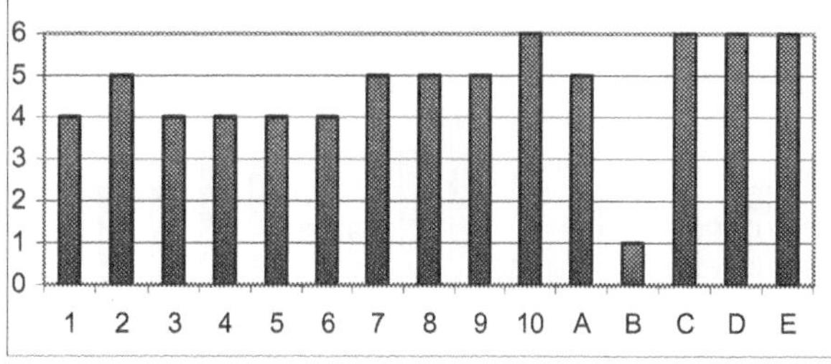

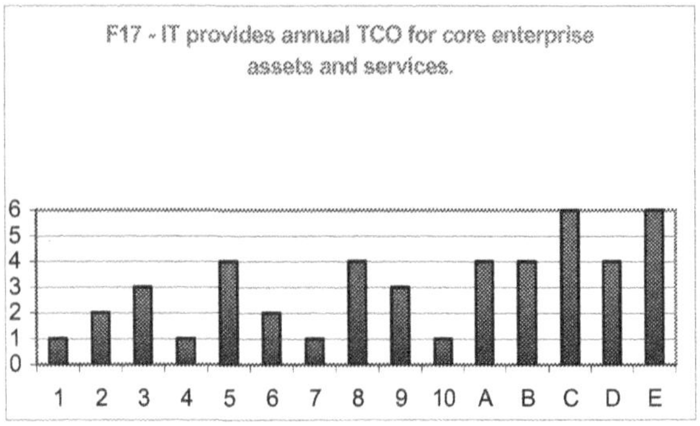
F17 - IT provides annual TCO for core enterprise assets and services.
0 1 2 3 4 5 6
1 2 3 4 5 6 7 8 9 10 A B C D E

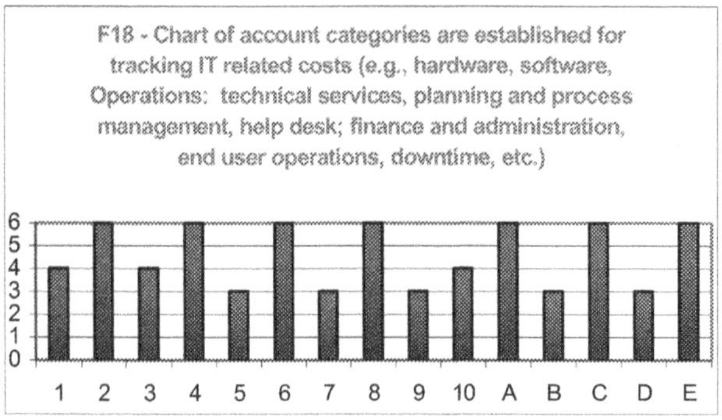
F18 - Chart of account categories are established for tracking IT related costs (e.g., hardware, software, Operations: technical services, planning and process management, help desk; finance and administration, end user operations, downtime, etc.)
0 1 2 3 4 5 6
1 2 3 4 5 6 7 8 9 10 A B C D E

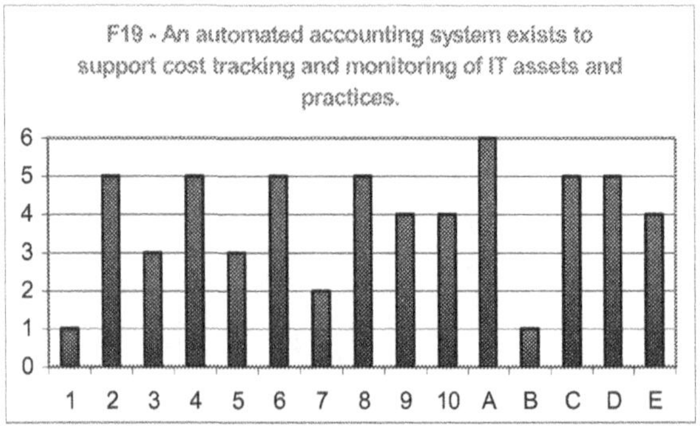
F19 - An automated accounting system exists to support cost tracking and monitoring of IT assets and practices.
0 1 2 3 4 5 6
1 2 3 4 5 6 7 8 9 10 A B C D E

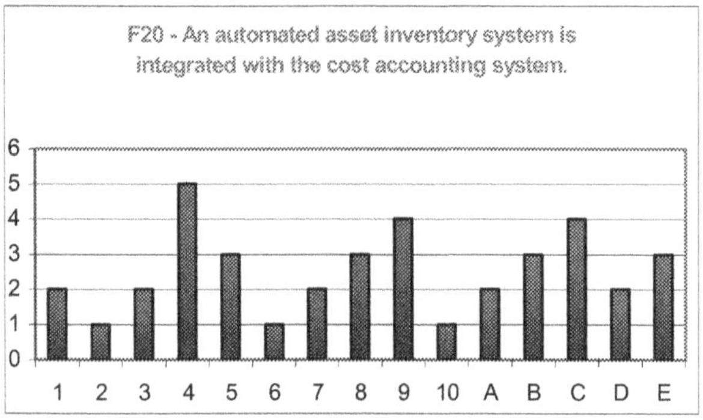
F20 - An automated asset inventory system is integrated with the cost accounting system.
0 1 2 3 4 5 6
1 2 3 4 5 6 7 8 9 10 A B C D E

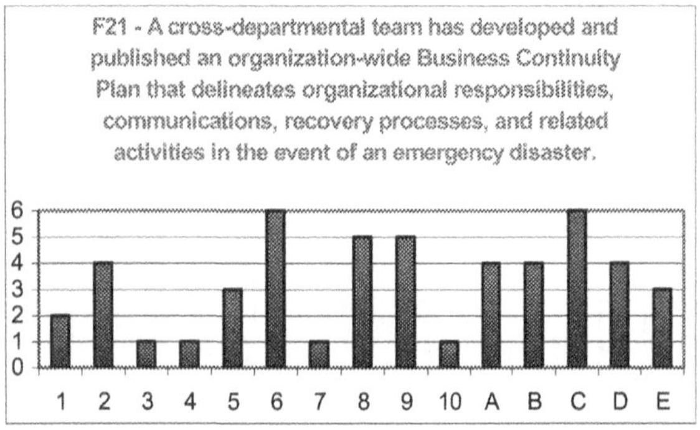
F21 - A cross-departmental team has developed and published an organization-wide Business Continuity Plan that delineates organizational responsibilities, communications, recovery processes, and related activities in the event of an emergency disaster.
0 1 2 3 4 5 6
1 2 3 4 5 6 7 8 9 10 A B C D E

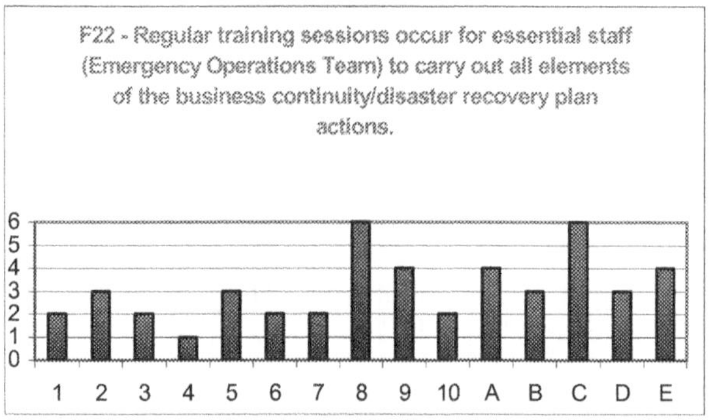
F22 - Regular training sessions occur for essential staff (Emergency Operations Team) to carry out all elements of the business continuity/disaster recovery plan actions.
0 1 2 3 4 5 6
1 2 3 4 5 6 7 8 9 10 A B C D E

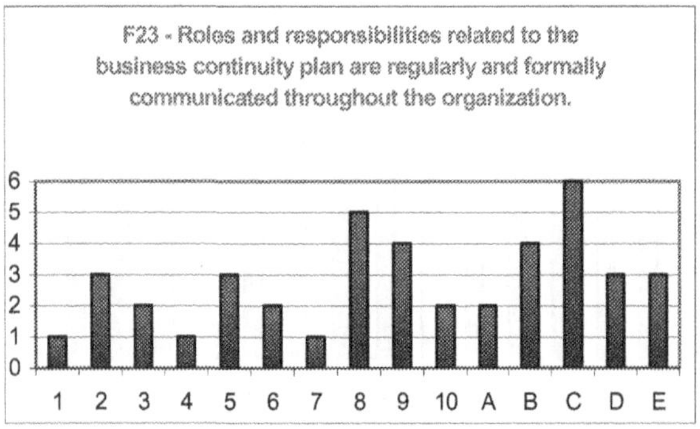
F23 - Roles and responsibilities related to the business continuity plan are regularly and formally communicated throughout the organization.
0 1 2 3 4 5 6
1 2 3 4 5 6 7 8 9 10 A B C D E

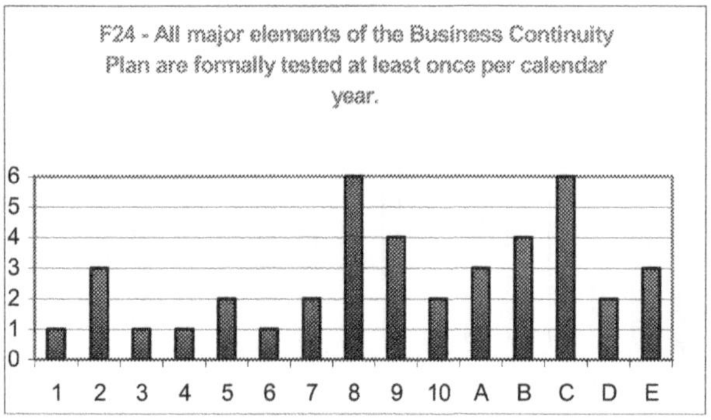
F24 - All major elements of the Business Continuity Plan are formally tested at least once per calendar year.
0 1 2 3 4 5 6
1 2 3 4 5 6 7 8 9 10 A B C D E

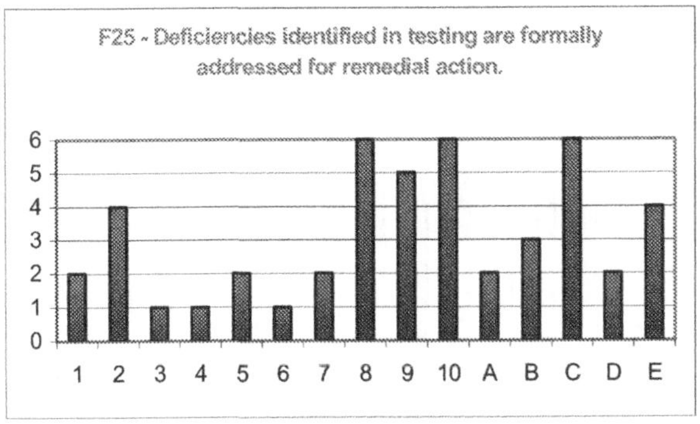


F25 - Deficiencies identified in testing are formally addressed for remedial action.

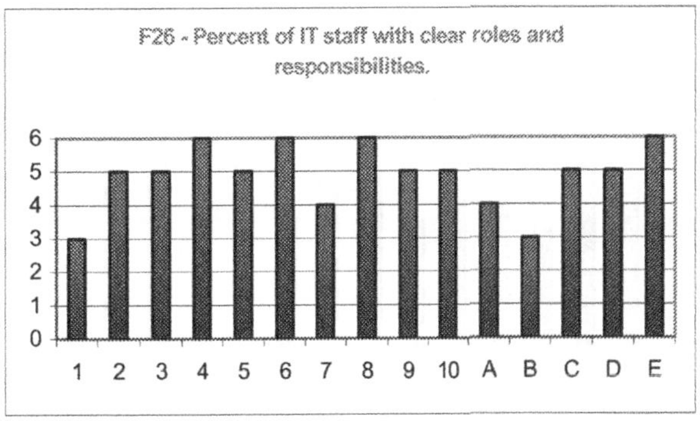


F26 - Percent of IT staff with clear roles and responsibilities.

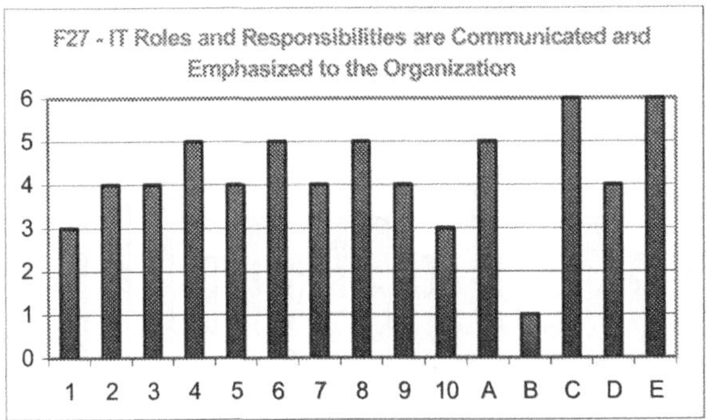


F27 - IT Roles and Responsibilities are Communicated and Emphasized to the Organization

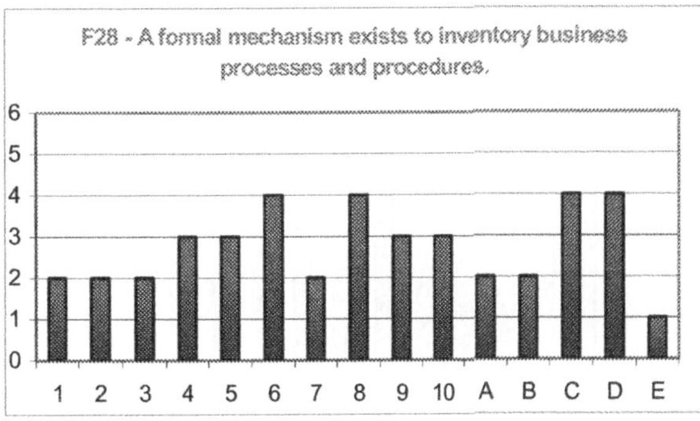


F28 - A formal mechanism exists to inventory business processes and procedures.

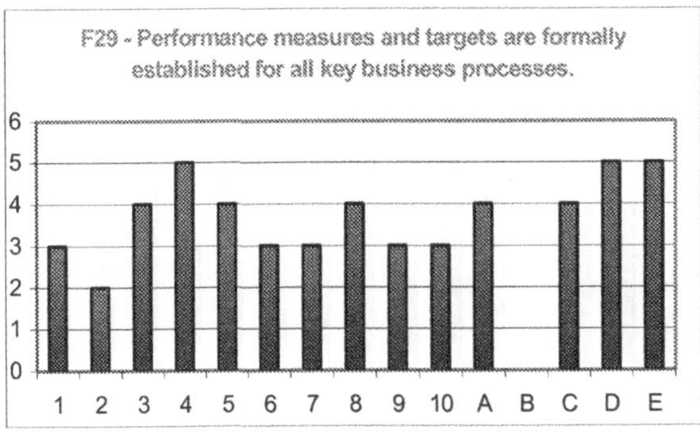


F29 - Performance measures and targets are formally established for all key business processes.

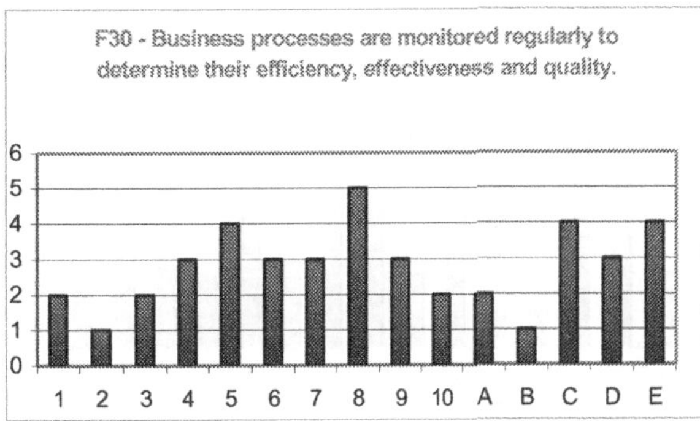


F30 - Business processes are monitored regularly to determine their efficiency, effectiveness and quality.

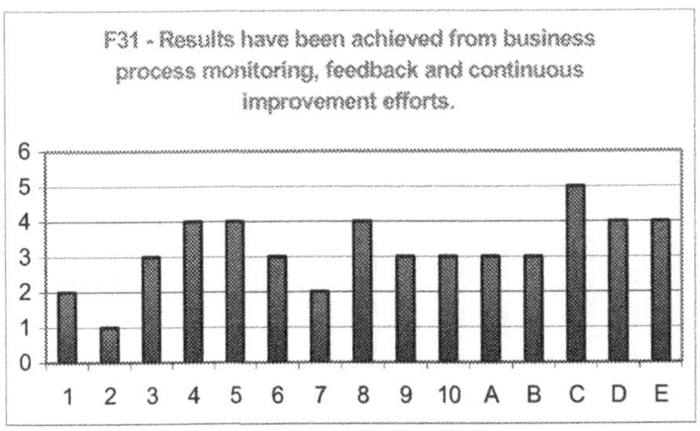


F31 - Results have been achieved from business process monitoring, feedback and continuous improvement efforts.

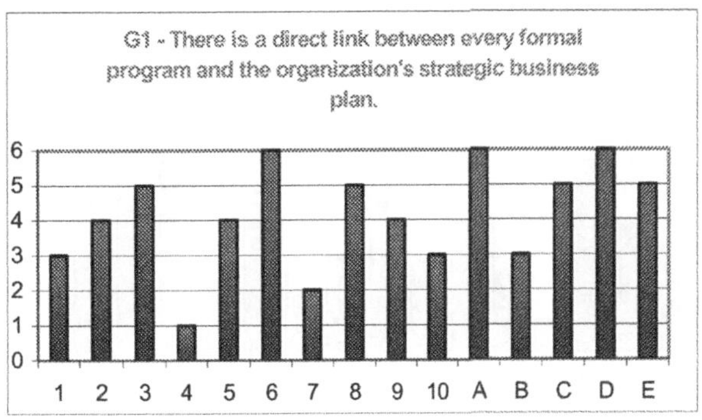


G1 - There is a direct link between every formal program and the organization's strategic business plan.

G2 - The annual IT budget reflects program priorities.

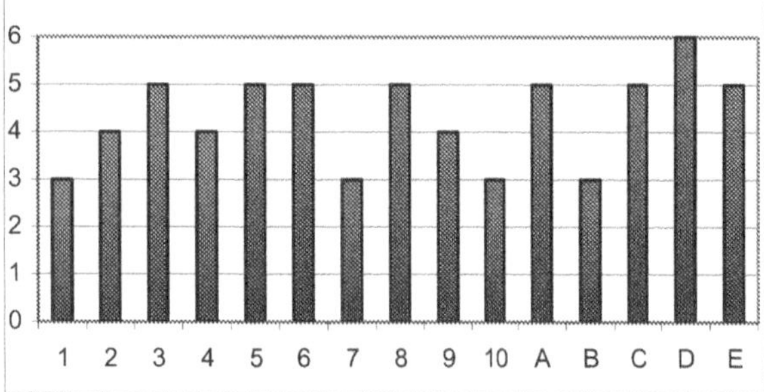


G3 - Every program is actively sponsored by one or more senior management representatives.

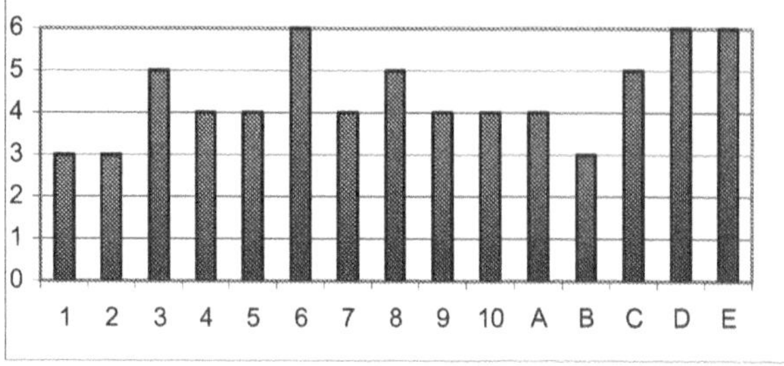


G4 - A Cross-Functional Group is Responsible for Coordinating Program Efforts.

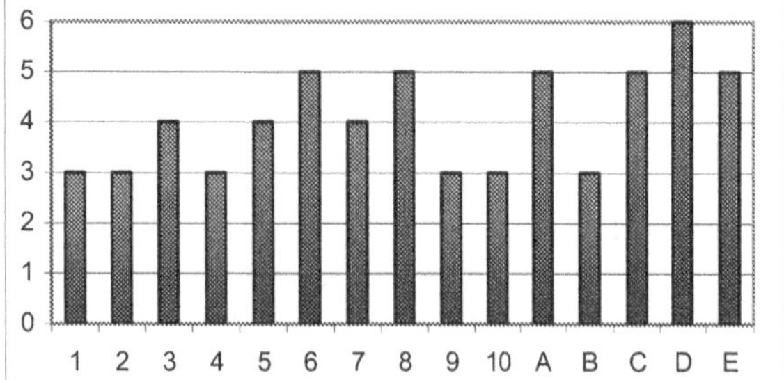


G5 - The programs expected outcome is clearly defined with objective measures, with progress tracked and regularly communicated.

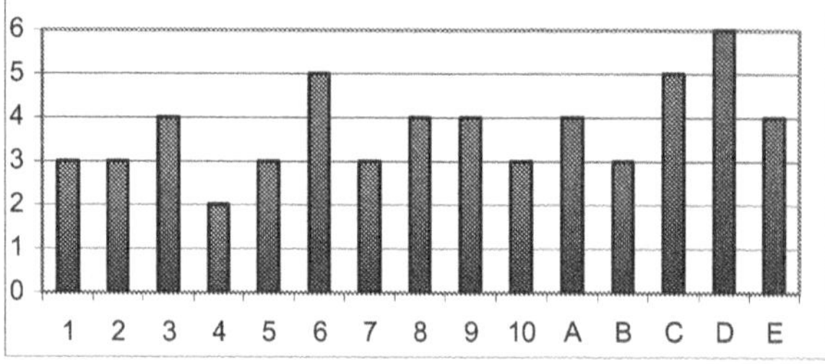


G6 - Project planning is aligned with the IT strategic plan.

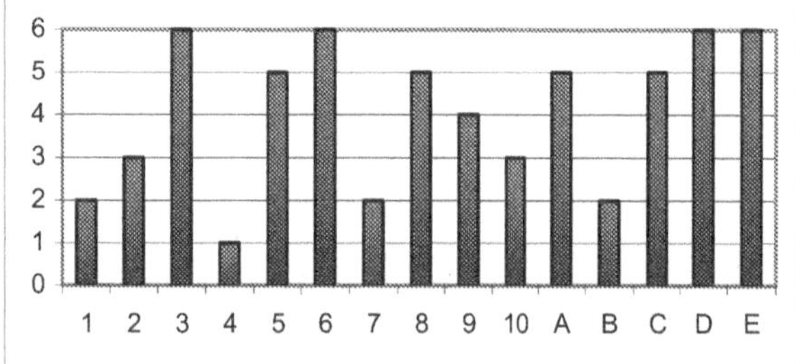


G7 - A formal, structured approach is used to review and approve project plans.

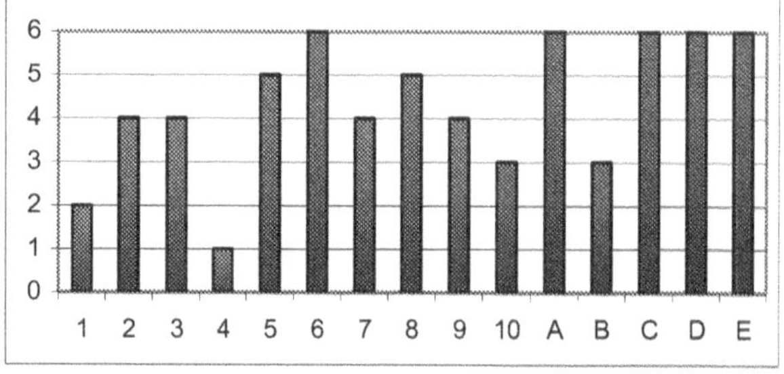


G8 - A project planning methodology has been adopted and is widely practiced.

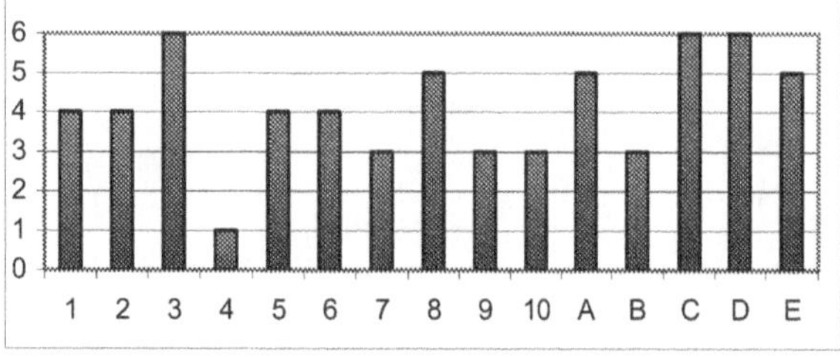


G9 - All projects have a plan with clear, traceable work breakdown structures, estimates, skill requirements, a quality plan and a change process.

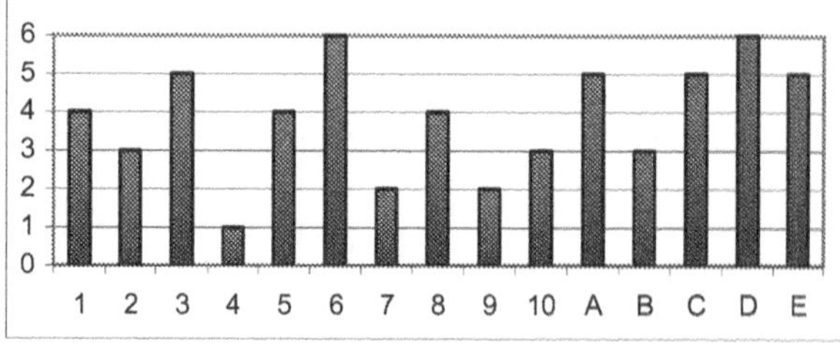

G10 - A scope statement is used to create agreement between the project team and the customer regarding project objectives and major project deliverables.

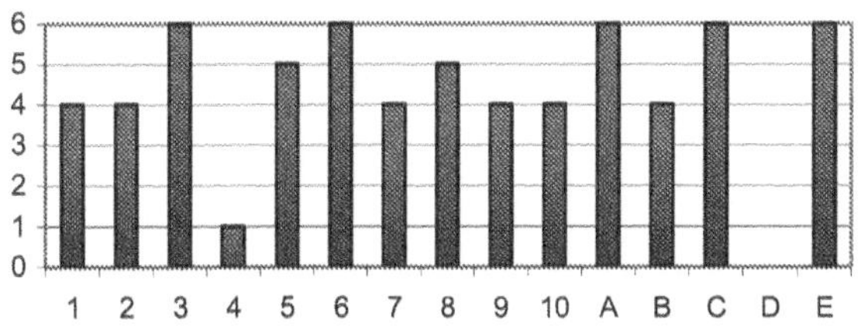


G11 - The number of projects completed on time, on budget, and with increased organization satisfaction with project delivered services, has increased.

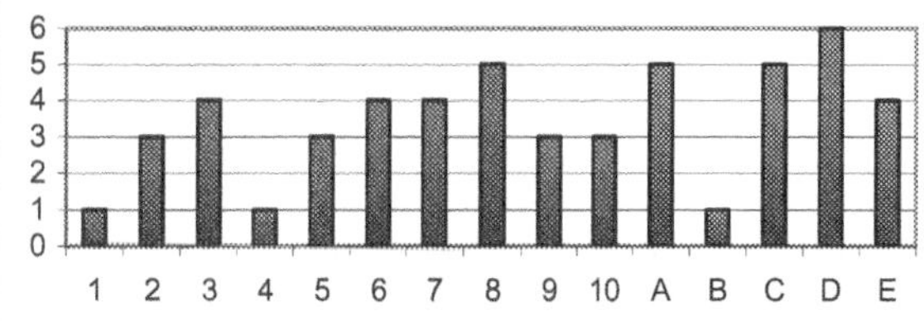


G12 - A formal quality assurance program and procedures exist; internal control, compliance and internal audit responsibilities are clearly understood.

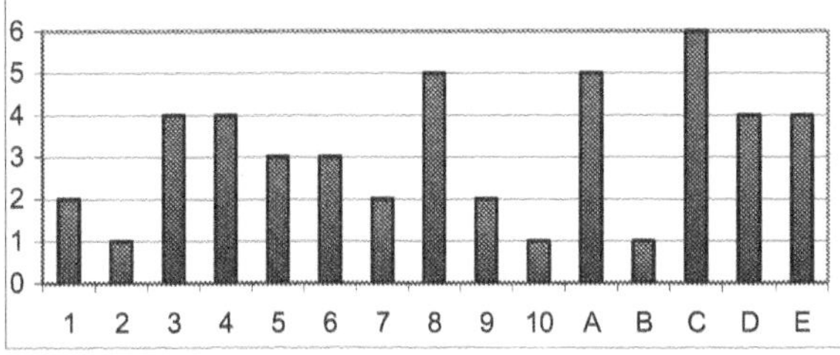


G13 - A formal process exists for tracking, monitoring, and trending QA program incidents and solutions.

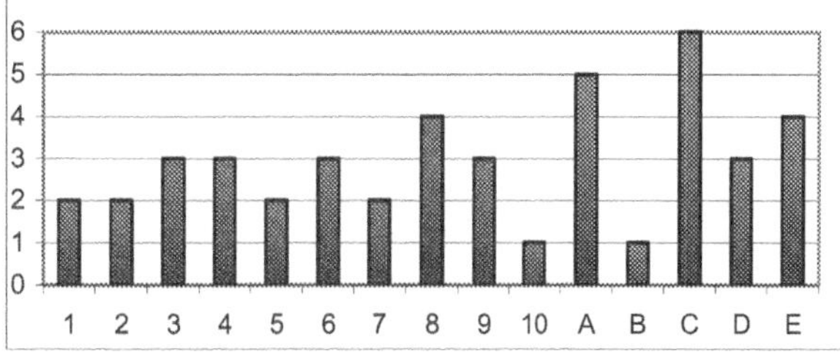


G14 - Formal quality control triggers are established, including inspection activities such as measuring, visual examination or testing to determine whether results conform to requirements.

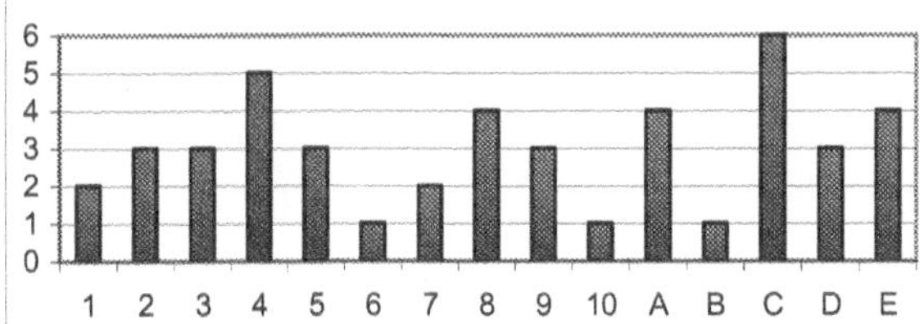


G15 - We have reduced IT quality problems by tracking the number of quality problems (defects) and non-compliance reports against quality standards.

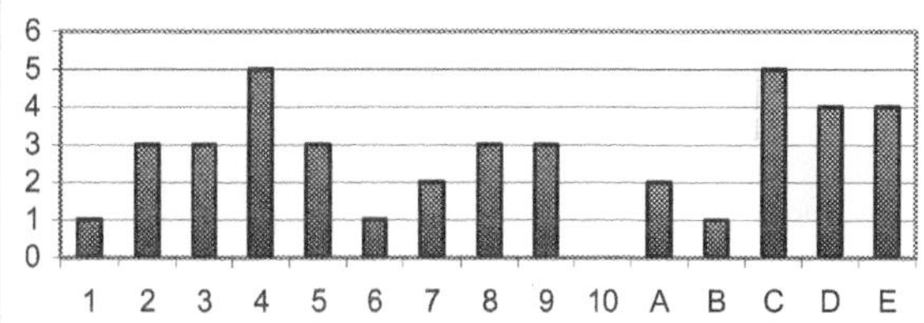


G16 - A formal risk assessment methodology is defined and enforced, including identifying, analyzing, and responding to project risks.

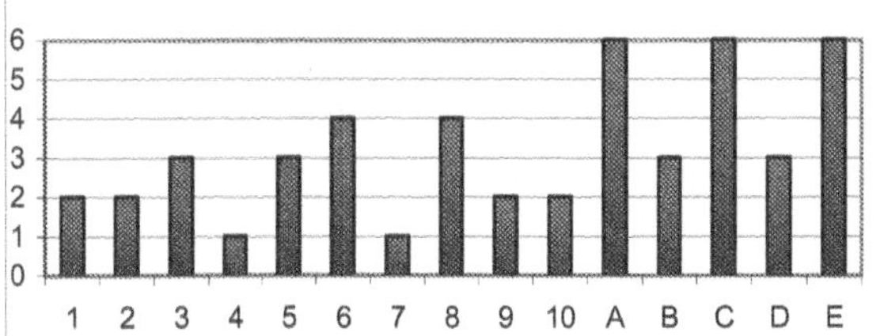


G17 - Project plans formally identify risks including measuring and managing value and risk prior to, during, and after project completion.

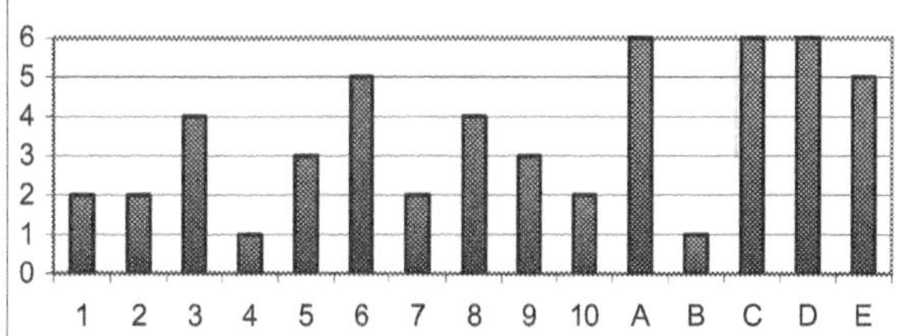

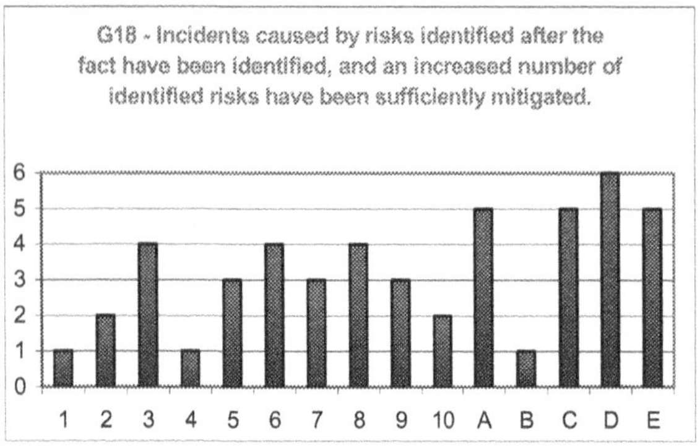
G18 - Incidents caused by risks identified after the fact have been identified, and an increased number of identified risks have been sufficiently mitigated.
0
1
2
3
4
5
6
1 2 3 4 5 6 7 8 9 10 A B C D E

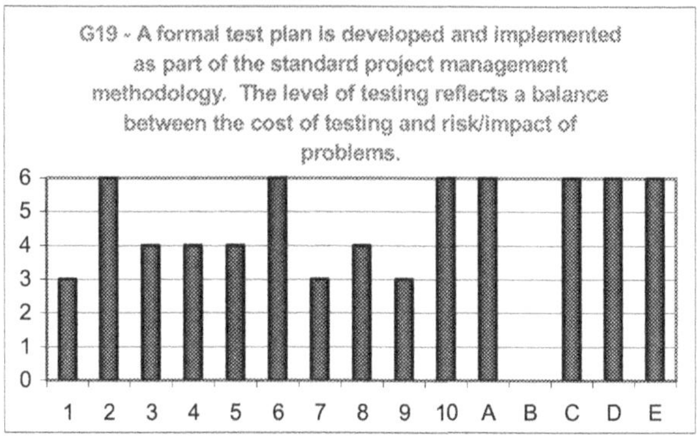
G19 - A formal test plan is developed and implemented as part of the standard project management methodology. The level of testing reflects a balance between the cost of testing and risk/impact of problems.
0
1
2
3
4
5
6
1 2 3 4 5 6 7 8 9 10 A B C D E

## SECURITY RESPONSES

Note that the order of these responses is ranked from lowest to highest to preserve the anonymity of the responses. Some questions that were included in the original survey were pertinent and repeated in this survey for clarity as a check. In these cases, the number of the original survey question is included in parentheses on the graphs.

In general, the responses were strikingly similar when utilities were asked the same questions in two different surveys; occasionally they responded slightly less positively when the question was solely in the context of security.

H.1. A list of known IT system vulnerabilities, system flaws, or weaknesses that could be exploited by threats has been developed and remedial actions have been taken to address these deficiencies.

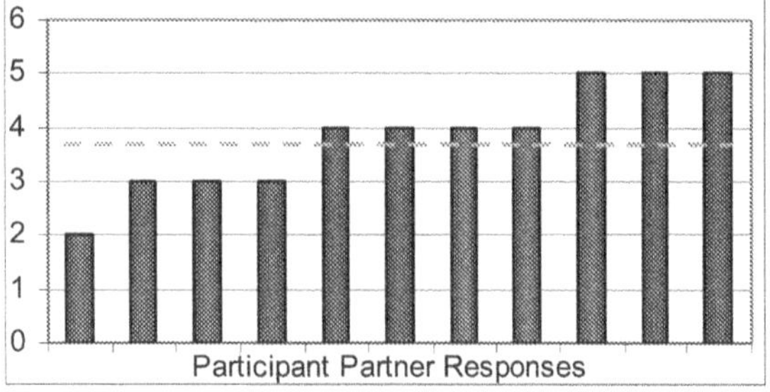


H.2. Analyses are conducted that determine whether in-place security processes adequately mitigate vulnerabilities.

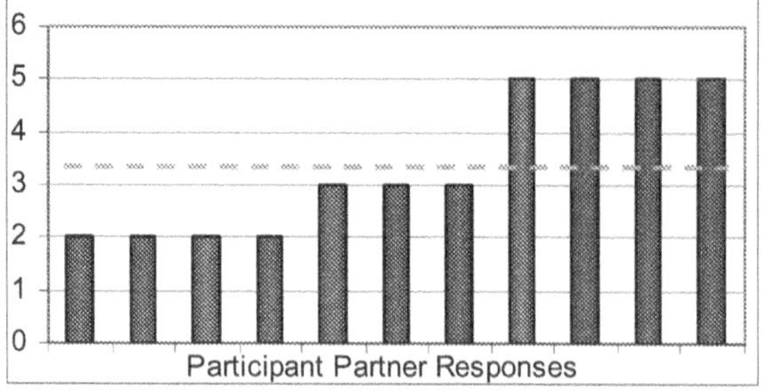


H.3. Industry-accepted password practices are in place.

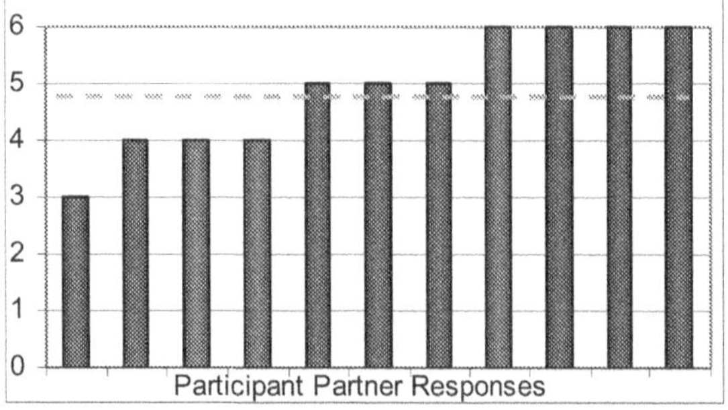


H.4. A process exists to identify, analyze and take remedial actions (when appropriate) for security alerts and/or incidents, which has shown to be effective in mitigating incidents or alerts.

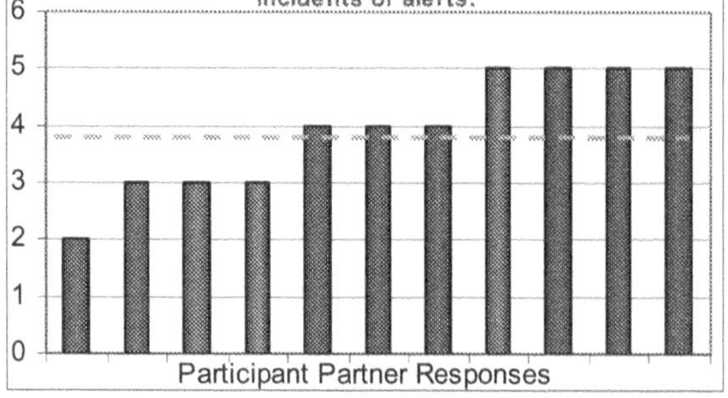


H.5. Tests and examinations of key controls are routinely performed (I.e., network scans, analyses of router and switch settings, penetration testing).

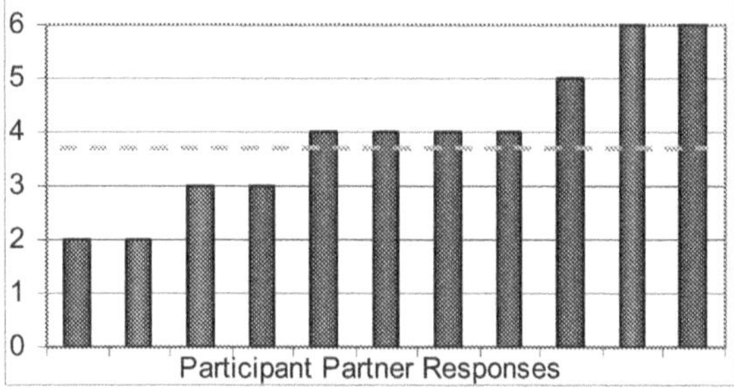


H.6. Operational, production-oriented systems (SCADA, process control systems) are reviewed for security, availability, reliability, and appropriate access to critical operation and production information.

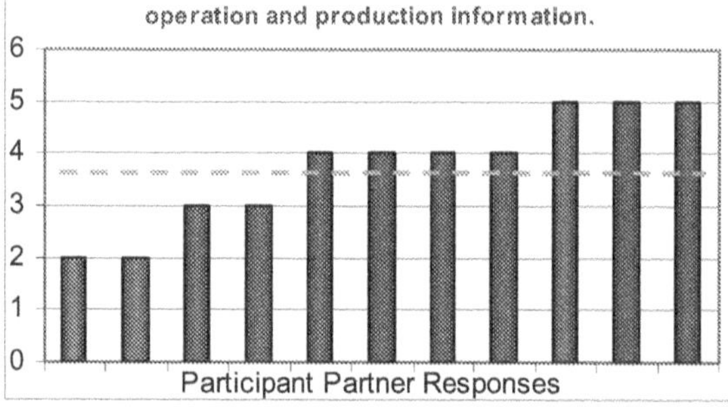


H.7. A process exists for background screening of appropriate personnel prior to granting systems access and authorization to bypass technical and operational controls.

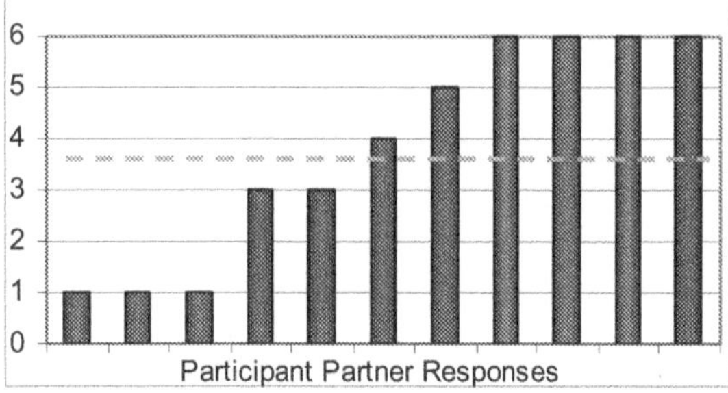


H.8. Confidentiality and security agreements are required for employees or contractors assigned to work with sensitive information.

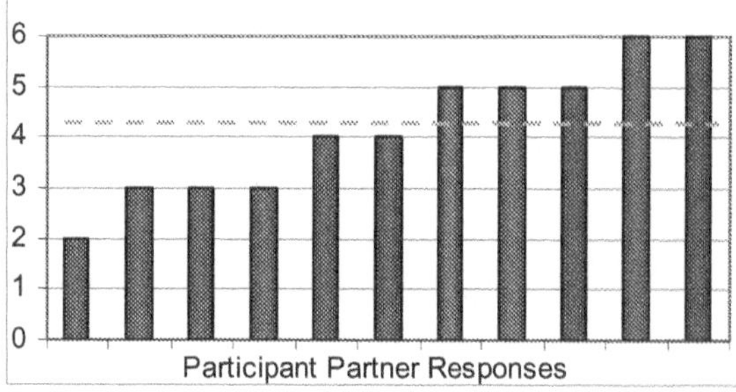

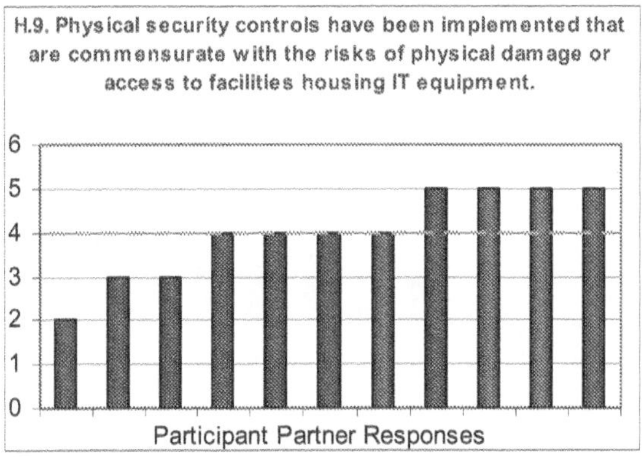
H.9. Physical security controls have been implemented that are commensurate with the risks of physical damage or access to facilities housing IT equipment.
6
5
4
3
2
1
0
Participant Partner Responses

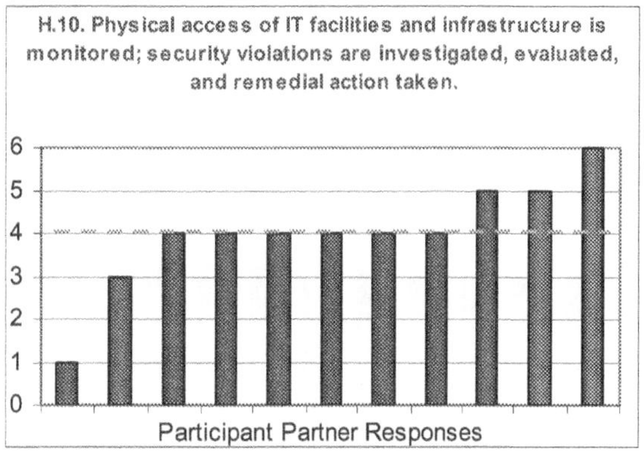
H.10. Physical access of IT facilities and infrastructure is monitored; security violations are investigated, evaluated, and remedial action taken.
6
5
4
3
2
1
0
Participant Partner Responses

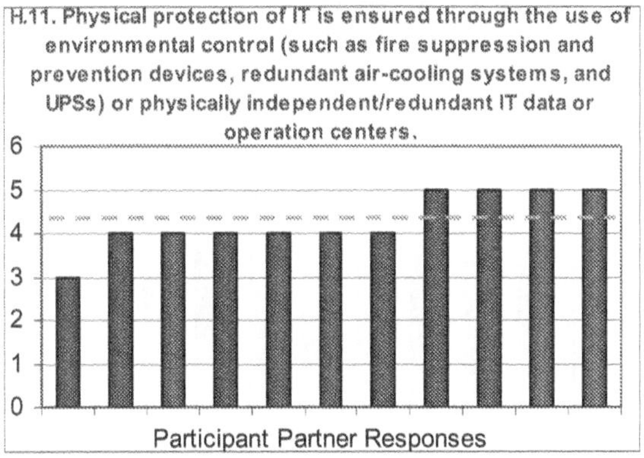
H.11. Physical protection of IT is ensured through the use of environmental control (such as fire suppression and prevention devices, redundant air-cooling systems, and UPSs) or physically independent/redundant IT data or operation centers.
6
5
4
3
2
1
0
Participant Partner Responses

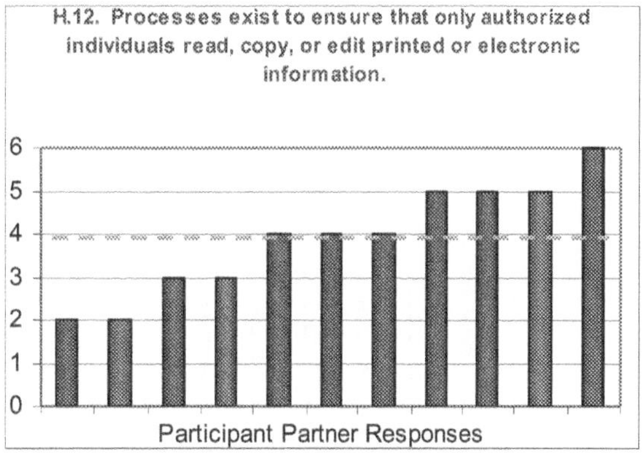
H.12. Processes exist to ensure that only authorized individuals read, copy, or edit printed or electronic information.
6
5
4
3
2
1
0
Participant Partner Responses

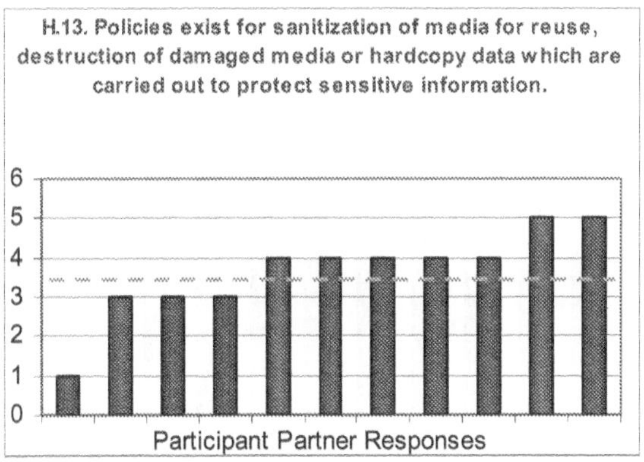
H.13. Policies exist for sanitization of media for reuse, destruction of damaged media or hardcopy data which are carried out to protect sensitive information.
6
5
4
3
2
1
0
Participant Partner Responses

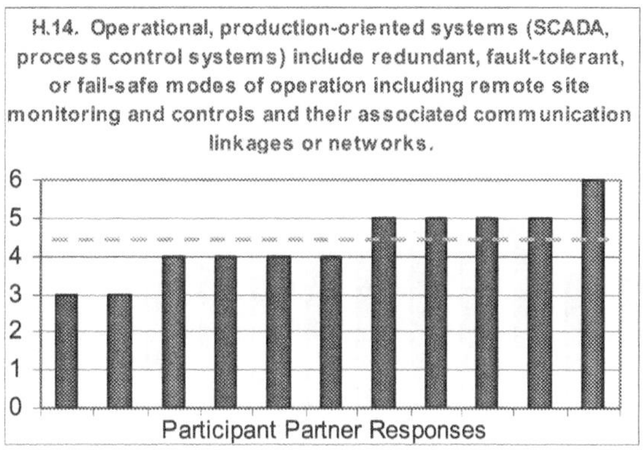
H.14. Operational, production-oriented systems (SCADA, process control systems) include redundant, fault-tolerant, or fail-safe modes of operation including remote site monitoring and controls and their associated communication linkages or networks.
6
5
4
3
2
1
0
Participant Partner Responses

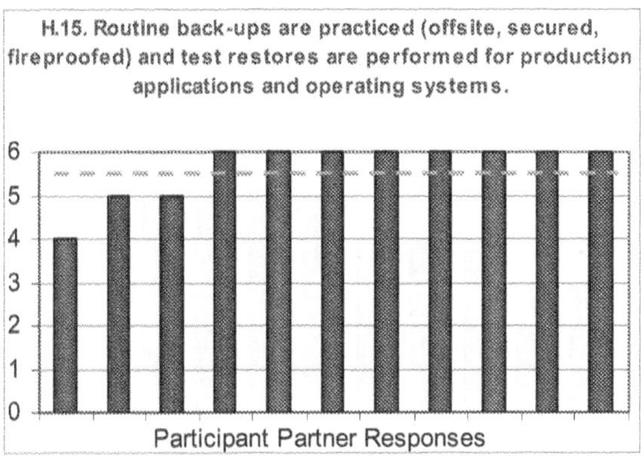
H.15. Routine back-ups are practiced (offsite, secured, fireproofed) and test restores are performed for production applications and operating systems.
6
5
4
3
2
1
0
Participant Partner Responses

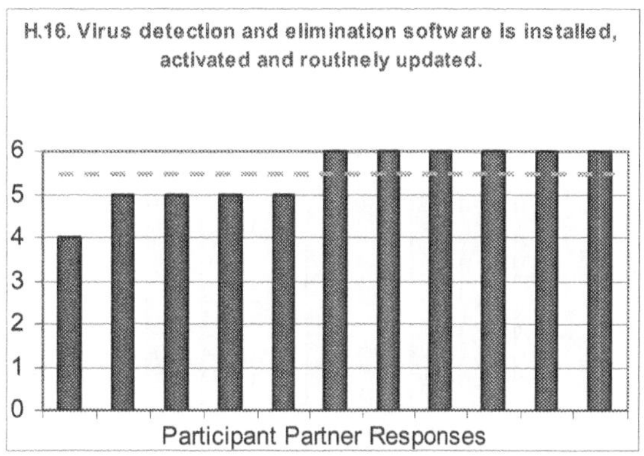
H.16. Virus detection and elimination software is installed, activated and routinely updated.
6
5
4
3
2
1
0
Participant Partner Responses

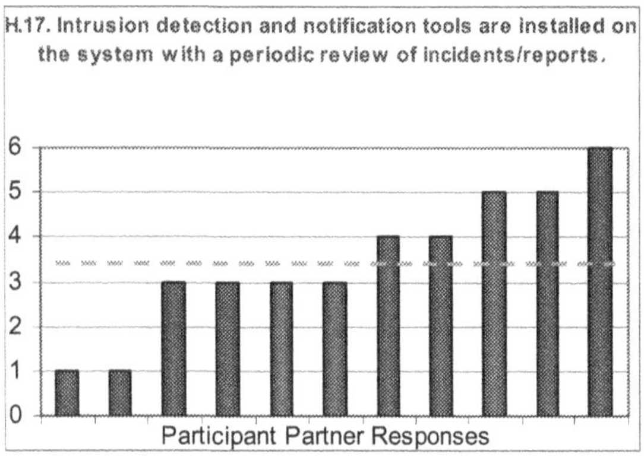
H.17. Intrusion detection and notification tools are installed on the system with a periodic review of incidents/reports.
6
5
4
3
2
1
0
Participant Partner Responses

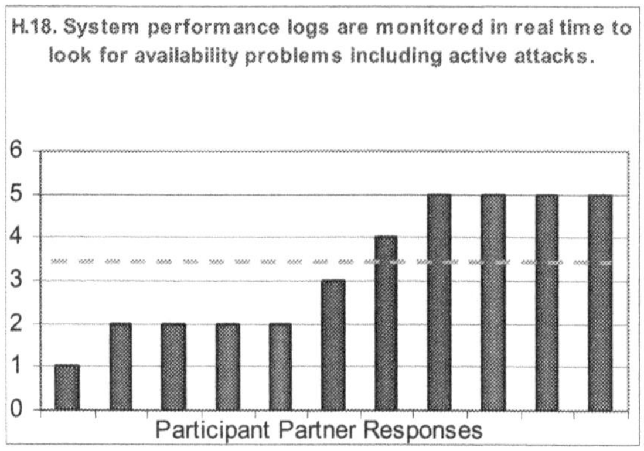
H.18. System performance logs are monitored in real time to look for availability problems including active attacks.
6
5
4
3
2
1
0
Participant Partner Responses

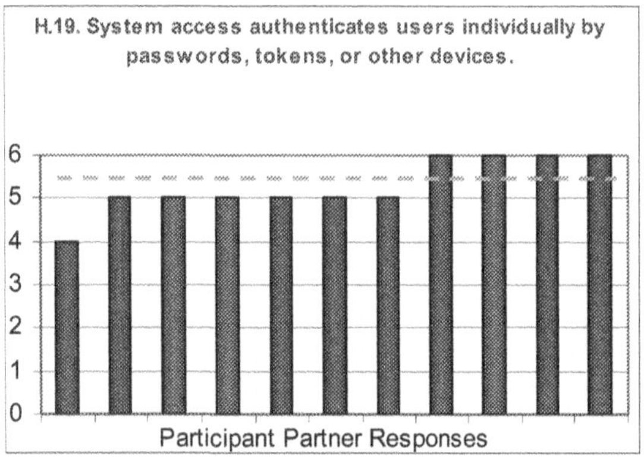
H.19. System access authenticates users individually by passwords, tokens, or other devices.
6
5
4
3
2
1
0
Participant Partner Responses

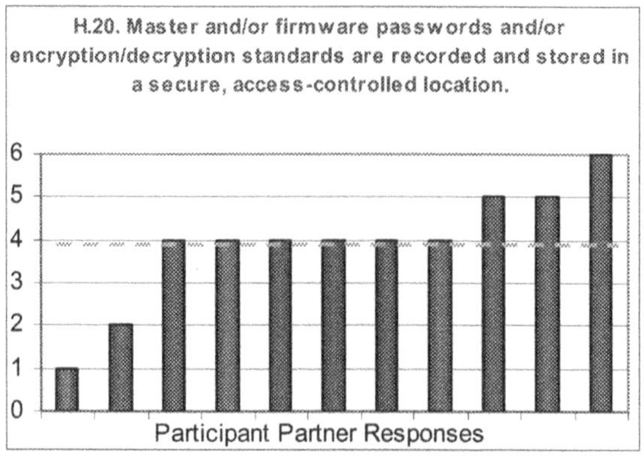
H.20. Master and/or firmware passwords and/or encryption/decryption standards are recorded and stored in a secure, access-controlled location.
6
5
4
3
2
1
0
Participant Partner Responses

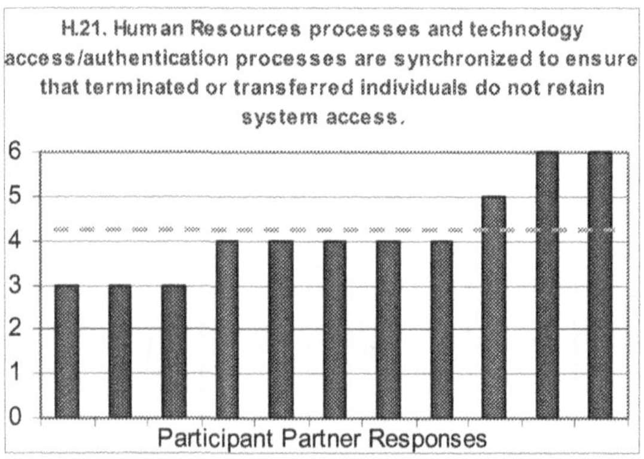
H.21. Human Resources processes and technology access/authentication processes are synchronized to ensure that terminated or transferred individuals do not retain system access.
6
5
4
3
2
1
0
Participant Partner Responses

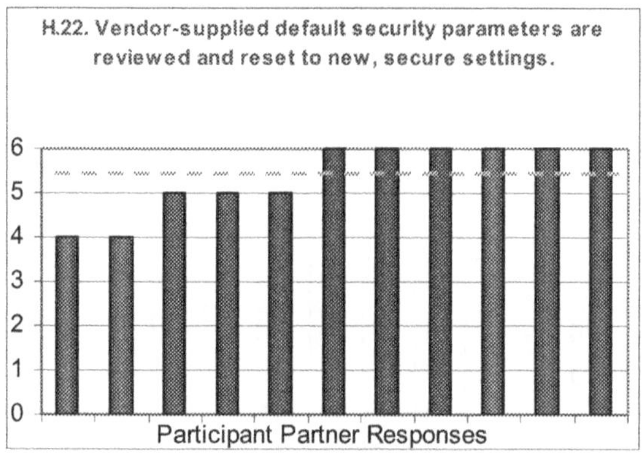
H.22. Vendor-supplied default security parameters are reviewed and reset to new, secure settings.
6
5
4
3
2
1
0
Participant Partner Responses

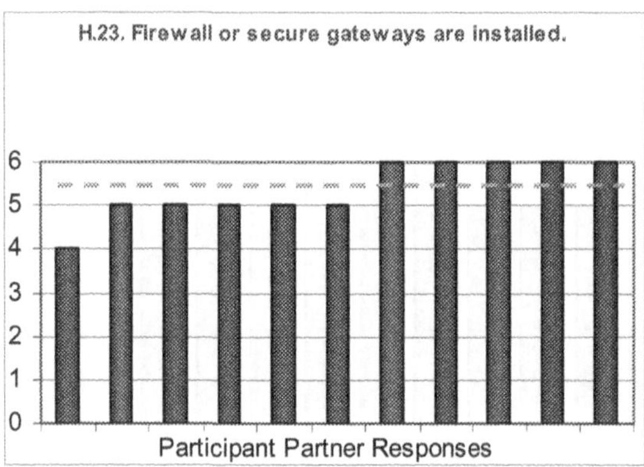
H.23. Firewall or secure gateways are installed.
6
5
4
3
2
1
0
Participant Partner Responses

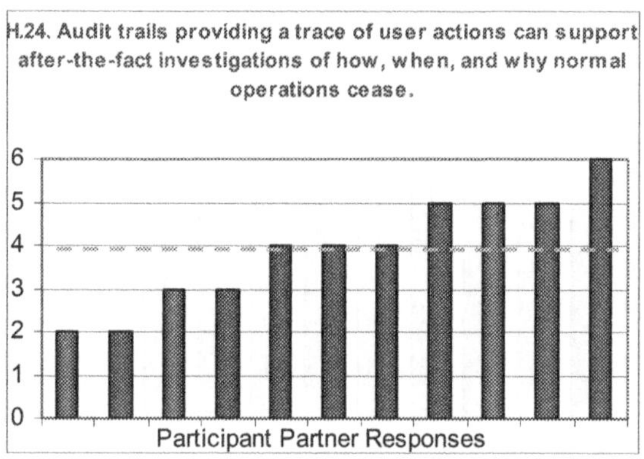
H.24. Audit trails providing a trace of user actions can support after-the-fact investigations of how, when, and why normal operations cease.
6
5
4
3
2
1
0
Participant Partner Responses

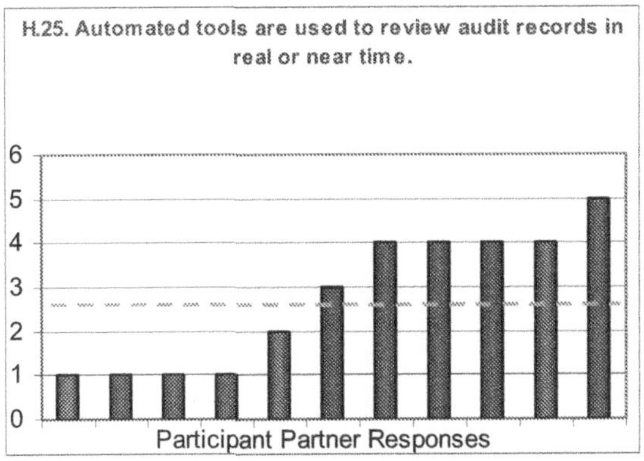
H.25. Automated tools are used to review audit records in real or near time.
6
5
4
3
2
1
0
Participant Partner Responses

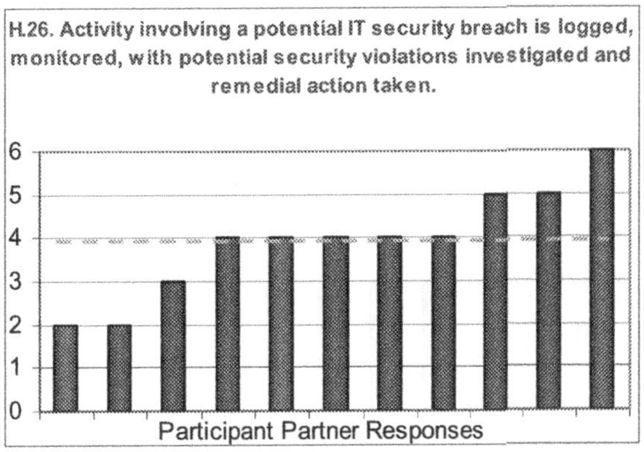
H.26. Activity involving a potential IT security breach is logged, monitored, with potential security violations investigated and remedial action taken.
6
5
4
3
2
1
0
Participant Partner Responses

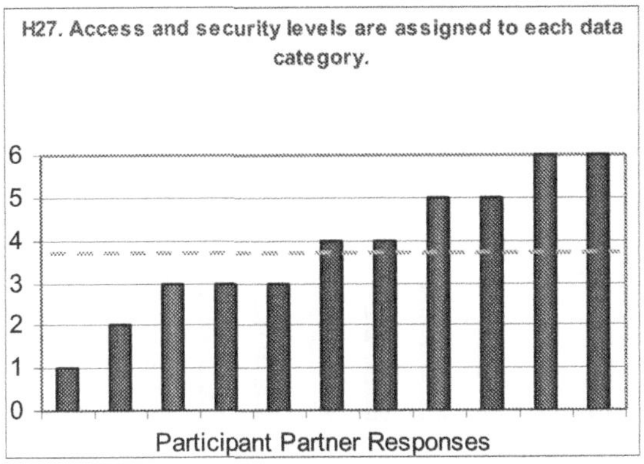
H27. Access and security levels are assigned to each data category.
6
5
4
3
2
1
0
Participant Partner Responses

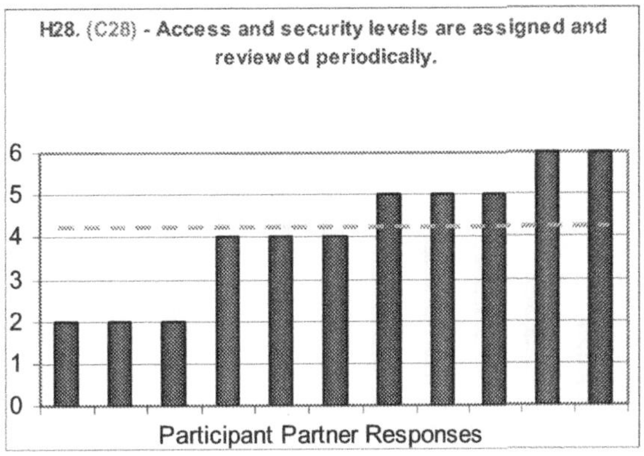
H28. (C28) - Access and security levels are assigned and reviewed periodically.
6
5
4
3
2
1
0
Participant Partner Responses

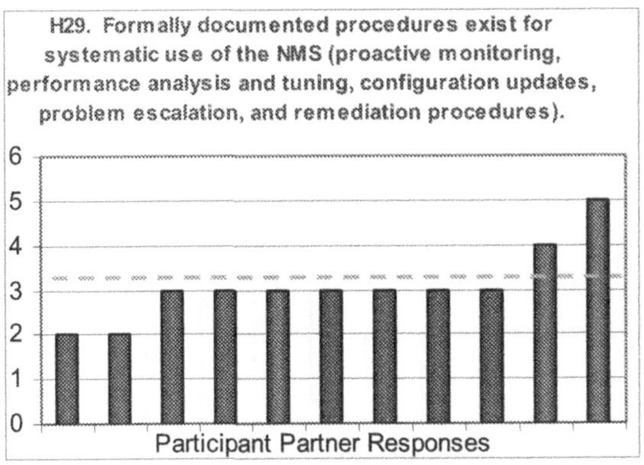
H29. Formally documented procedures exist for systematic use of the NMS (proactive monitoring, performance analysis and tuning, configuration updates, problem escalation, and remediation procedures).
6
5
4
3
2
1
0
Participant Partner Responses

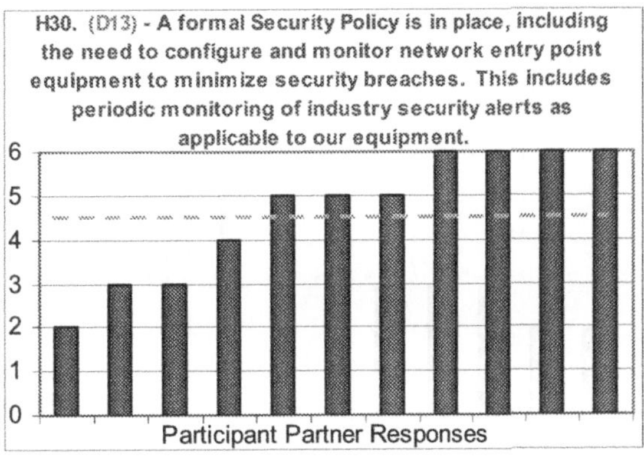
H30. (D13) - A formal Security Policy is in place, including the need to configure and monitor network entry point equipment to minimize security breaches. This includes periodic monitoring of industry security alerts as applicable to our equipment.
6
5
4
3
2
1
0
Participant Partner Responses

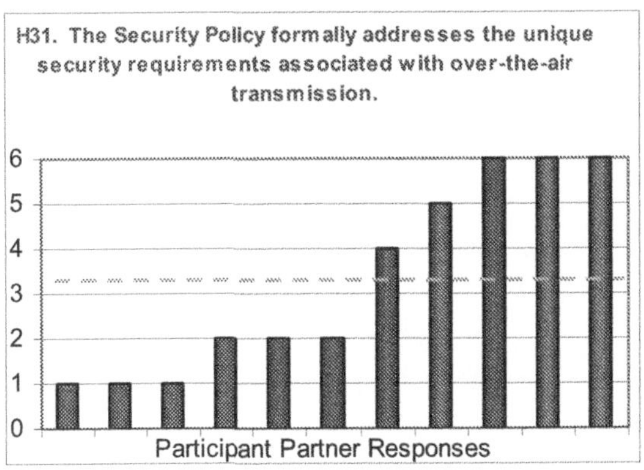
H31. The Security Policy formally addresses the unique security requirements associated with over-the-air transmission.
6
5
4
3
2
1
0
Participant Partner Responses

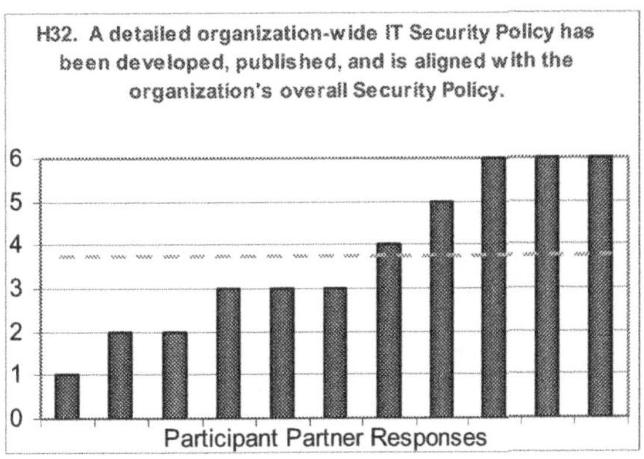
H32. A detailed organization-wide IT Security Policy has been developed, published, and is aligned with the organization's overall Security Policy.
6
5
4
3
2
1
0
Participant Partner Responses

H33. ( D22) - All employees receive formal education and training on the IT Security Policy.

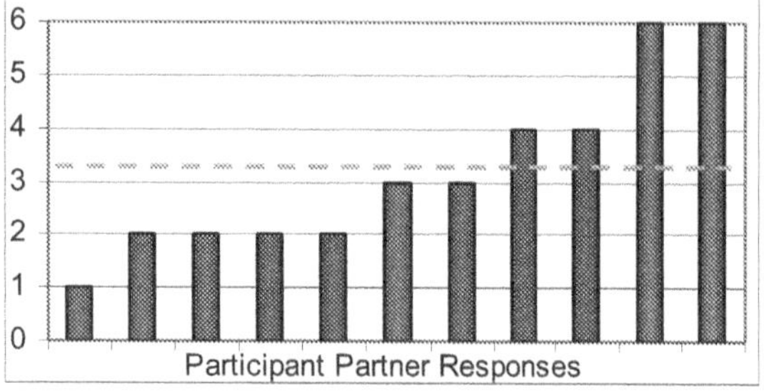


H34. Adequate IT resources are allocated to implement the IT security policy.

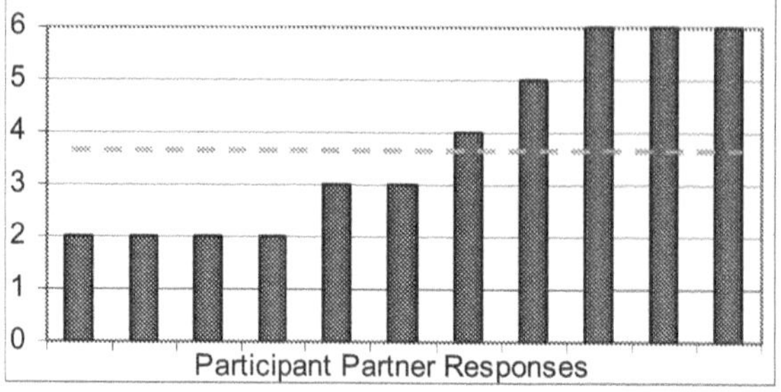


H35. (F21) - A cross-departmental team has developed and published an organization-wide Business Continuity Plan that delineates organizational responsibilities, communications, recovery processes, and related activities in the event of an emergency disaster.

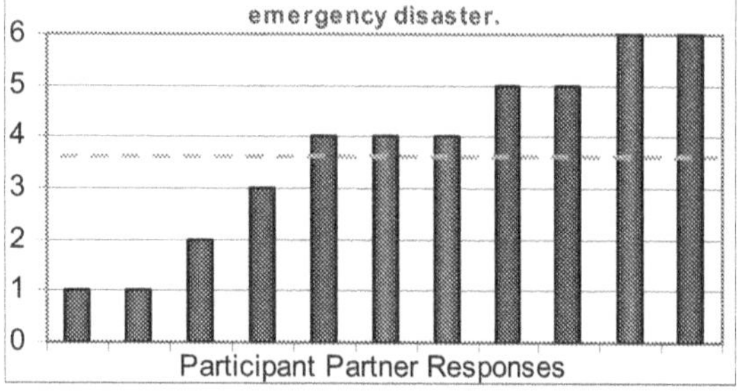


H36. (F22) - Regular training sessions occur for essential staff (Emergency Operations Team) to carry out all elements of the business continuity/disaster recovery plan actions.

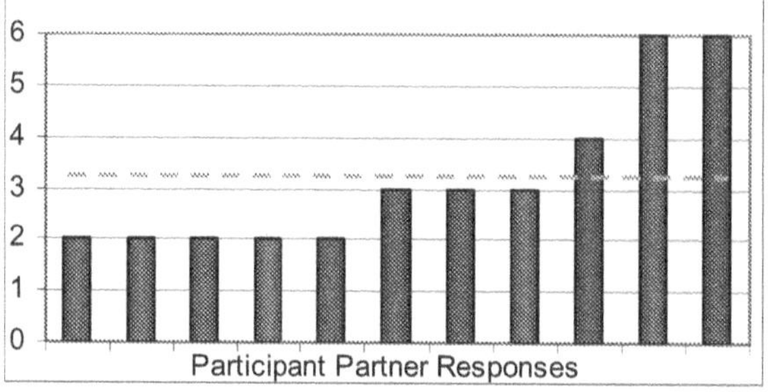


H37. (F23) - Roles and responsibilities related to the business continuity plan are regularly and formally communicated throughout the organization.

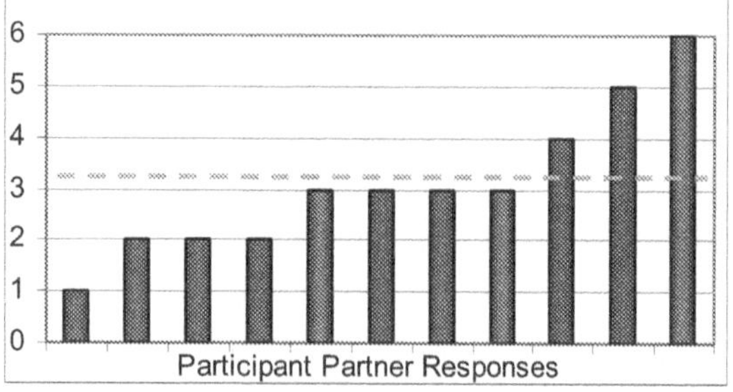


H38. (F24) - All major elements of the Business Continuity Plan are formally tested at least once per calendar year.

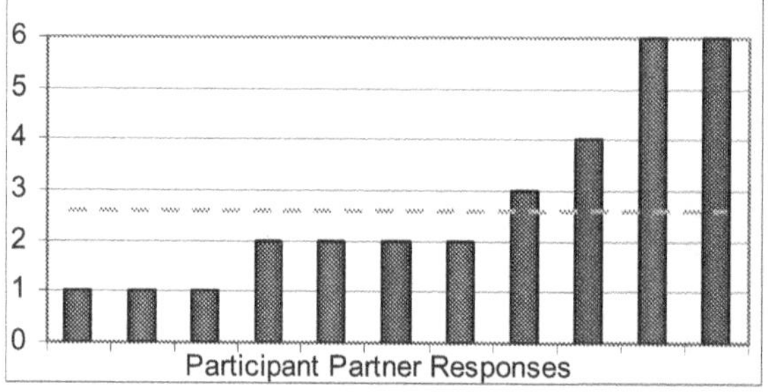

# APPENDIX C
# PARTICIPATING UTILITY CONTRIBUTED PRACTICES

Participating utilities were asked if they would like to contribute unique or leading practices that might be of interest to other utilities, as a part of this project. A number of practices were contributed, as identified below, and described on the following pages:

| **Utility** | **Interesting or Leading Practice** |
|---|---|
| Cleveland | Teaching IT project management to non-IT staff |
| East Bay | Integrating databases (water quality, GIS/MAPS, customer information) |
| EPCOR | SCADA Integration (Providing real-time and historical information to all business units) |
| Honolulu | IT Telephony |
| Honolulu | Defining and implementing IT staffing and development/roles and responsibilities (including broadband compensation) |
| Phoenix | Implementing standards for Process Control Systems within other applications |
| San Diego | EAI Enterprise Application Integration (applying middleware tools and processes) |
| San Diego | Implementing data management, quality and stewardship policies (LIMS Application and GIS) |
| Tucson | Implementing a Permit Tracking Database |

| ORGANIZATION | FOCUS AREA | PRACTICE |
|---|---|---|
| **Cleveland Water** | **G – Program and Project Management** | **Project Management** |

### 1. Practice Description

This practice explains how we use project management to help leverage the success of IT projects.

### 2. How Achieved – Time Required to Achieve

- We provide project management classes, with the classes taught by a professor at a local college, to employees who agree to act as future project managers for our IT projects. The classes are given on city time, at a cost of $1,200 per student. Each course lasts from six (6) to nine (9) weeks (depending on the class size), for a full day (eight hours) per week. Upon passing the class, employees are eligible to apply for the Project Management Professional certification test.
- This program has been in place for about three years. Part of the coursework includes developing and presenting a project to the rest of the class. Generally students select real projects in which they are involved, for their case studies.
- No additional pay is received unless the students need to work overtime. They must commit to be a project manager on one of our projects in order to take the class, and their supervisor must approve of their participation. If the student wants to join the Project Management Institute (PMI) we pay for the PMI membership as well as the test fee, test preparation materials, and classes. Thus far about 50% of the employees trained on this program have decided to join the PMI.
- We assign the trained project managers to all IT projects under the direction of our internal Project Management Office (PMO). We currently have five full-time people in the PMO; they supervise the newly trained PMs. Each major project has two PMs assigned – one from the PMO (Lead PM) and the other one of the students (Deputy PM). One of these two must be a subject matter expert. Pairing the PMs provides the new PMs with confidence, and provides a support system for the two project managers.

### 3. Benefits Resulting From Practice

- The non-IT PMs provide different insights into the projects since they come from the user and business support base. Most of them are mid-managers and degreed people who have not been able to really exercise their degree in their job. This program has created very strong enthusiasm among the students, since they have actively chosen to get involved.
- We are using this system for all of our significant IT projects (LIMS, security assessment, financial management/budget module, WMS, document management system, and HR). In addition, our project managers are being used for the GIS and telecommunications city-wide endeavors.

### 4. Why the Organization Feels it is a Leading Practice

- We have seen the benefit of using strict project management techniques and have been able to build a large combination of trained employees.
- We completed our first project using this approach – this project (replacing our financial management/budget module) came in one month early, and 3% under the budget.

### 5. Metrics Used to Ascertain Effectiveness

Projects completed on-time and on-budget.

| 6. Areas for Practice Improvement |
|---|
| We would like to see more of our trained staff seek certification. |
| **7. Related Survey Practice and Metric Questions** |
| G8, G9, G11 |

| ORGANIZATION | FOCUS AREA | PRACTICE |
|---|---|---|
| **EBMUD** | **Applications & Integration** | **Application Development** |
| **1. Practice Description** | | |
| EBMUD engineers rely heavily on their hydraulic model to assist in making planning decisions for their water and recycled water systems. In the early to mid-90s, it was becoming very cumbersome for the many engineers (approximately thirty to forty engineers) to use their system, which was not integrated, and didn't support multiple users. In addition, using and configuring the existing model took significant training on the part of junior engineers. Once trained, the engineers often moved to other areas such as design or construction – thus triggering the need to train new engineers. So EBMUD surveyed existing off-the-shelf hydraulic modeling products in search of a package that would integrate GIS, CIS, facility data, and hydraulic modeling itself. At the time, none existed. Thus the District embarked upon a project to develop it's own Hydraulic Modeling System. | | |
| **2. How Achieved – Time Required to Achieve** | | |
| This project was driven by Engineering, with eventual collaboration between the IS and Engineering Departments. Engineering is still responsible for maintaining the applications. EBMUD also consulted with outside industry to ascertain future directions. We have two team members dedicated to the maintenance and upgrade of the application. A user's group has been established to re-evaluate effectiveness and usefulness of applications with individuals responsible for benchmarking the application against what is available off the shelf. The total time from concept to development took about 2-3 years. The tool has since gone through two major revisions, as EBMUD's requirements have grown and technology has developed. | | |
| **3. Benefits Resulting From Practice** | | |
| Engineering staff is able to accurately model the distribution system for various situations including seismic upgrades, system upgrades, outage plans, system improvements, future improvements, and so on. Many one-time analyses are able to use the same data for different purposes. We only have to maintain the data one time. Additionally, Engineering has major buy-in to the application, because they had so much involvement in developing and maintaining it. Because we have the skills in-house, there has been a reduced need for outsourcing. Expertise is maintained in-house. | | |
| **4. Why the Organization Feels it is a Leading Practice** | | |
| Our District was able to achieve this in-house years before the consultant industry was able to develop and offer a like product. Our system is already customized to our specific needs; expertise is maintained in-house and training is conducted in-house. We were fortunate that we were large enough to justify the investment. | | |
| **5. Metrics Used to Ascertain Effectiveness** | | |
| We have made significant reductions in the amount of time required for planning projects. For example, the time required to provide plans for water service has been reduced by approximately four weeks. There has been a major reduction in the time required for us to provide estimates and develop plans. | | |

| • 6. Areas for Practice Improvement |
|---|
| Evaluation of new developments in GIS and other off-the-shelf hydraulic modeling software may make this application obsolete or just provide ideas for further enhancements. |
| **7. Related Survey Practice and Metric Questions** |
| B.1C |

| ORGANIZATION | FOCUS AREA | PRACTICE |
|---|---|---|
| **Honolulu Board of Water Supply** | **F: IT Organization and Direction** | **Roles and Responsibilities** |

| 1. Practice Description |
|---|
| Reorganized the IT department to better respond to the needs of the Honolulu Board of Water Supply (HBWS.) Introducing "broad banding" in an attempt to compensate employees for their increased productivity and expertise. |
| **2. How Achieved – Time Required to Achieve** |
| They created and staffed a special projects section for GIS and other related projects and also reorganized the commercial and operations staff into an Applications group and an Operational Support group. The groups are all matrix oriented, with any team member acting as a project manager and development teams being formed as necessary. An effort is underway to "broadband" IT staff into three categories instead of the current 20. This will allow them to use union staff in any capacity needed, get the training they need, compensate them properly for work performed, and expand their duties and responsibilities. Time required to achieve is 4 months. |
| **3. Benefits Resulting From Practice** |
| Applications and operations special project groups have formed teams and completed many successful projects since the reorganization. Teams are formed as needed and project management skills are pushed to everyone. New personnel have become identified as leaders and they now have more options to take on additional projects. |
| **4. Why the Organization Feels it is a Leading Practice** |
| The old structure was designed to support legacy systems. All personnel were tied to the "CAS group" or the "Payroll group." Little development took place, and little or no knowledge transfer took place. The new groups have learned new disciplines, new skills, and have been given career paths. Morale is much higher and much more work is getting done. |
| **5. Metrics Used to Ascertain Effectiveness** |
| Amount of projects completed and in work. |
| **6. Areas for Practice Improvement** |
| Would like to push forward with "Broad banding" and eventually be able to properly compensate IT staff for the work performed. |

| 7. Related Survey Practice and Metric Questions |
|---|
| ▪ F.26, F.27 |

| ORGANIZATION | FOCUS AREA | PRACTICE |
|---|---|---|
| **Honolulu Board of Water Supply** | **D: It Infrastructure and Management** | **Telephony** |

**1. Practice Description**

This element addresses the convergence of traditional telephony technology with IT infrastructures and systems. It includes the functionality of PBXs (telephone switches) and adjunct systems such as Interactive Voice Response (IVR), Automatic Call Distribution (ACD), and Unified Messaging (voicemail, e-mail, video mail, etc.)

**2. How Achieved – Time Required to Achieve**

Key business drivers for moving to a new telephony system:

- 18 year old production PBX systems had limited functionality,
- Lacked further expandability
- Expensive to use and maintain
- Employees find it cumbersome to use
- Customers were being dropped or lost within the current phone system (call center).

They first develop an overall phone strategy to an integrate call center, dispatch, and call plan functions (IVR, call routing, auto attendant, etc.) They then used a telephony-consulting firm to drive detailed requirements and assist with recommended solutions and integration. They're currently in the process of implementing a new voice-over-IP (VoIP) solution with 450 CISCO phones being deployed within one business unit and the operations center. They are also assessing their network infrastructure to ensure there is no single point of failure within the voice/data network and that there is sufficient bandwidth to support the new voice-over-IP (VoIP).

**3. Benefits Resulting From Practice**

- Improved customer service with a centralized call center for routing and managing all customer calls.
- Reduced long-term cost by merging the voice network with the network.
- Unified messaging (voice, data, document management, and fax) can be managed within the Exchange email system. A single user interface (Exchange) can be used for managing multiple technologies and/or applications.

**4. Why the Organization Feels it is a Leading Practice**

The convergence of voice and data onto a single network helps make collaboration tools a reality and thus a leading practice.

| 5. Metrics Used to Ascertain Effectiveness |
|---|
| ROI over time (system will pay for itself in phone charges), additional options for staff, more flexibility and added capability for customer service call center, dispatch, and IT help desk. |
| **6. Areas for Practice Improvement** |
| Call routing, IVR, phone/system integration. Currently, many department managers and line supervisors names appear in the local phone book. Calls from outside the company are going to individual employees instead of a centralized call center. Many calls are lost or fall through the cracks. Known network problems (Ex: broken water main) are reported many times to many employees which creates confusion and delays with dispatch and repair, etc. |
| **7. Related Survey Practice and Metric Questions** |
| D.8, D.10 |

| ORGANIZATION | FOCUS AREA | PRACTICE |
|---|---|---|
| **EPCOR** | **B: Applications and Integration** | **Integration Tools and Processes** |
| **1. Practice Description** | | |
| The vision of a perfectly integrated application portfolio is far from today's reality, though many utilities are actively working toward the vision. The issues around integration include unique data models for each application, data collection timing differences, data granularity issues (e.g., varying levels of precision and detail (time periods, geography)), vendor approaches, different uses of the same data, and constantly changing technologies. Key to an integrated environment is an enterprise architecture approach. Technologies being developed and implemented to provide an integrated environment are typically complex and require high levels of sophistication. | | |
| **2. How Achieved – Time Required to Achieve** | | |
| The distinctiveness of EPCOR Water Services control systems from mainstream utilities is not so much the type of hardware and software being used, but how the systems are being utilized in helping with day-to-day operations and as well in long range planning. The EPCOR control systems have been highly integrated and can readily provide real-time and historical information (data) via a browser interface to anyone in the corporation. The initial implementation was completed within 18 months with continual improvements being made. | | |
| **3. Benefits Resulting From Practice** | | |
| Plant information has been made widely available and visible to not only the core plant operations personnel but to the business side as well (i.e. throughout the enterprise.) Enterprise wide availability of both real-time and historical data fosters sharing of information and collaboration, it has also improved communication and utilization of expertise across the utility. The essence of this is that the overall intelligence about the plant improves which can lead to quicker and more informed decision-making on how best to operate the plant. More importantly, the utilization of this information is extensive and increasing which helps with improving water quality, consistency of operation through conventional and advanced controls, cost control for chemical and energy usage. | | |

| 4. Why the Organization Feels it is a Leading Practice |
|---|
| Over the past decade, EPCOR Water Services has also adopted a strategy of making regular incremental improvements to the control systems through the yearly capital program as opposed to the traditional methodology still practiced by other utilities of making intermittent, massive and expensive upgrades once the systems become largely obsolete. This strategy has helped to ensure that the systems remain current, reliable, and cost effective. It is indeed also this strategy that has had a direct and significantly positive effect in the development of in-house expertise in the area of controls. It has been shown that this expertise is more advanced when compared to many automation consultants. |
| **5. Metrics Used to Ascertain Effectiveness** |
| A number of metrics or measures are used to determine Return On Investment (ROI) and overall effectiveness:<br>1) The ability to more quickly and easily integrate automation systems into its plants<br>2) How quickly they can integrate external plants for remote control and operation<br>3) To reduce the number of consultants working on controls related projects<br>4) The ability to provide competitive bids with higher water quality standards. |
| **6. Areas for Practice Improvement** |
| Would like to implement a more robust event and performance management monitoring tool to ensure that the network and applications have not been infiltrated. |
| **7. Related Survey Practice Questions** |
| B.32 |

# APPENDIX D
# HIGHLIGHTS OF GARTNER GROUP PRESENTATION

- Trends: Creating Your Effective Enterprise Architecture and e-Workforce Solutions
  - Organizations are defining Enterprise Architectures to leverage managed change
    - ***Assumption***: Organizations adopting an Enterprise Architecture approach will realize a 30 percent improvement in their operational ability to address constantly changing external drivers (0.6 probability)
  - Industry trends and the demand for IT labor are leading to an 'IS Lite" environment
    - Trend #1 – Sharper demarcation of IT activities leads to role clarity between central/local activities and governance

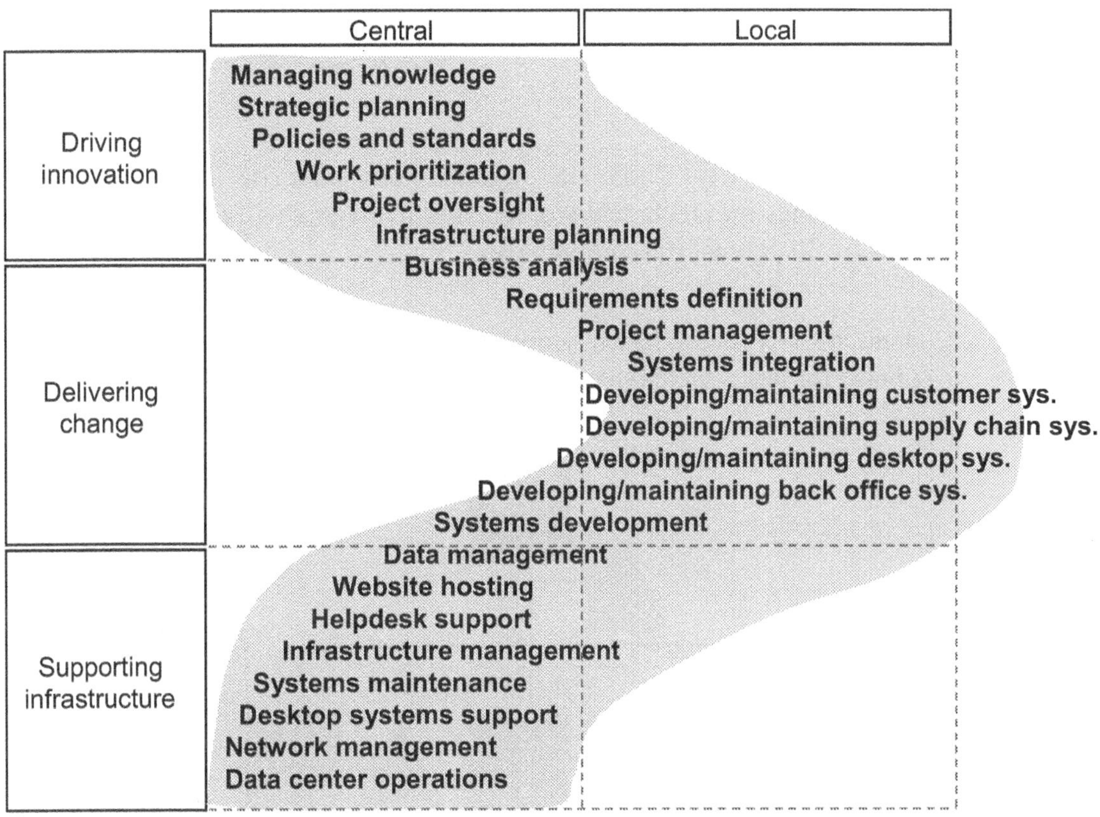

– Trend #2 – Specialization in centers of excellence leads to pooling and nurturing of specialist skills

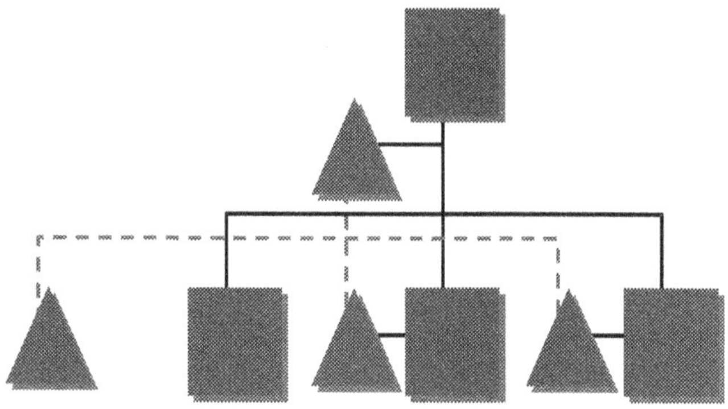

*What activities where, and who is responsible?*

- Resource pool
- Technology tracking
- Project office
- Relationship management
- Architecture/technical standards
- ERP
- Knowledge management
- IT purchasing
- Support for a business activity (e.g., customer service)

– ***Assumption***: By 2005, IS organizations that do not instill business competencies into technical workers and managers will miss business-defined goals by 30 percent (0.8 probability)

– Trend #3 – Process-based working leads to improved responsiveness and support for large-scale business change end-to-end service

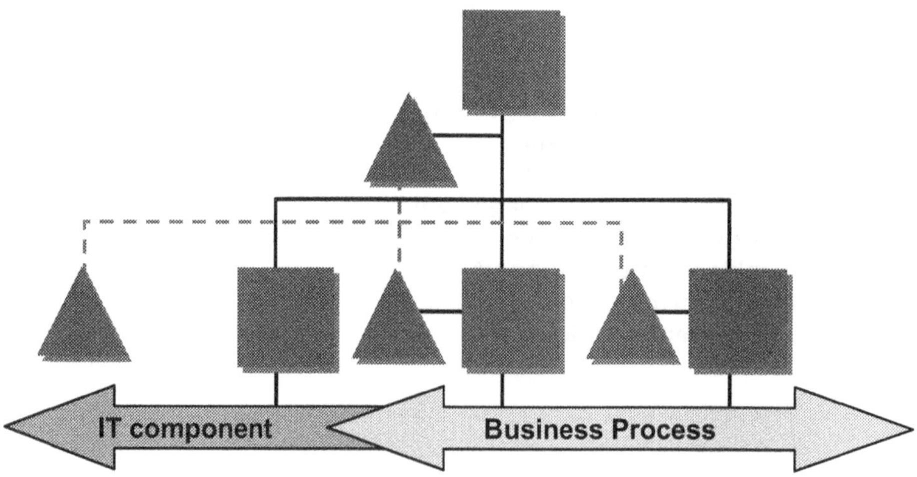

– Trend #4 – Outsourcing leads to leveraging vendor resources and focusing internal resource on higher-value work

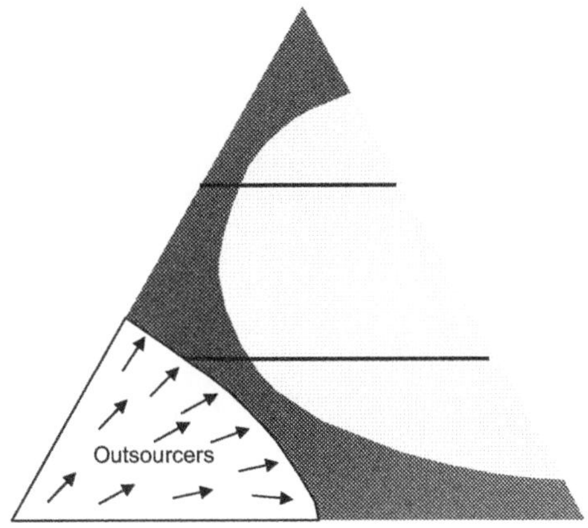


– ***Assumption***: By 2004, one-third of the current IT job roles will be replaced by new ones (0.8 probability)
– ***Assumption***: Through 2004, enterprises that focus on IT worker motivation will be twice as likely to attract and keep qualified people as those that focus only on retention (0.8 probability)
– ***Assumption***: By 2002, enterprises that align management style with worker's fulfilling mission-critical job roles will face half the turnover within those job roles (0.7 probability)

– Five key roles are emerging
  - Vendor Management
  - Technology Advancement
  - Business Enhancement
  - Architecture Development
  - IT Leadership

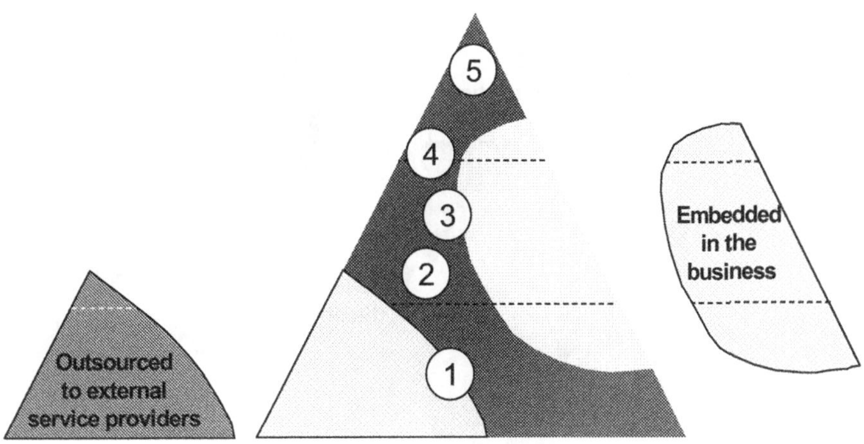

– Future IT roles will require greater business and behavioral competencies – not just technical

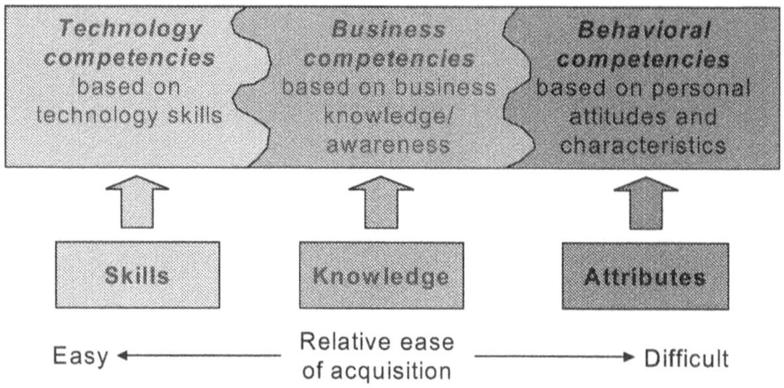


Rather than hire on the basis of technical ability, hire based on behavioral competencies

– Training and development help to bridge gaps – but match the training approach to the need
– Offer the widest range of rewards you can – but tailor them to the needs of each individual

| | |
|---|---|
| CRITICAL | 1. Challenging Work<br>2. (Market-based) Pay-Performance Link<br>3. Credible Leaders |
| HIGH PRIORITY | 4. Flexibility & Trust<br>5. Training/Career Development<br>6. Ownership<br>7. Communication<br>8. Creative Freedom<br>9. Pleasing Work Environment<br>10. Flexible Benefits |

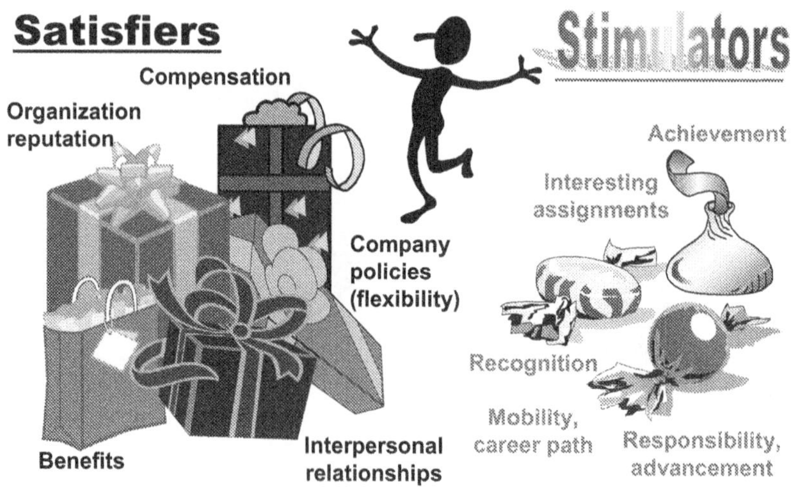

***Gartner Top 10 Predictions for IT in 2002:***

- *External Forces*
  - The IT industry will remain challenged, facing accelerated job losses and significant vendor consolidation
  - Safeguarding people, knowledge, systems and nations will take priority
  - Consumers will go online, finally, with the number using online account management doubling by 2005
- *Business Behavior*
  - Short-term focus on expenses will squeeze IS organizations in 2002 as business demand for IT increases
  - Outsourcing and trusted suppliers will take more control as capital spending reduces in favor of operating budgets
  - Through 2004, businesses will continue to view the discipline of CRM as a critical component of corporate strategy
- *Applications and Technology Trends*
  - During 2002, despite budget restraints, operational IT infrastructure will still need to anticipate and fulfill critical IT initiatives
  - More than 50 percent of mobile applications deployed at the start of 2002 will be obsolete by the end of 2002
  - By 2004, Web services will dominate deployment of new application solutions for Fortune 2000 companies
  - During 2002, leading-edge businesses will exploit application integration to generate business innovation

# APPENDIX E
# INFORMATION TECHNOLOGY FRAMEWORK

The framework is comprised of eight focus areas, each described more fully below:

- Applications and Integration
- Data
- Infrastructure Management
- Service Delivery and Sourcing
- IT Organization Direction/Oversight and Planning
- IT Security
- Program and Project Management
- Overall/Common

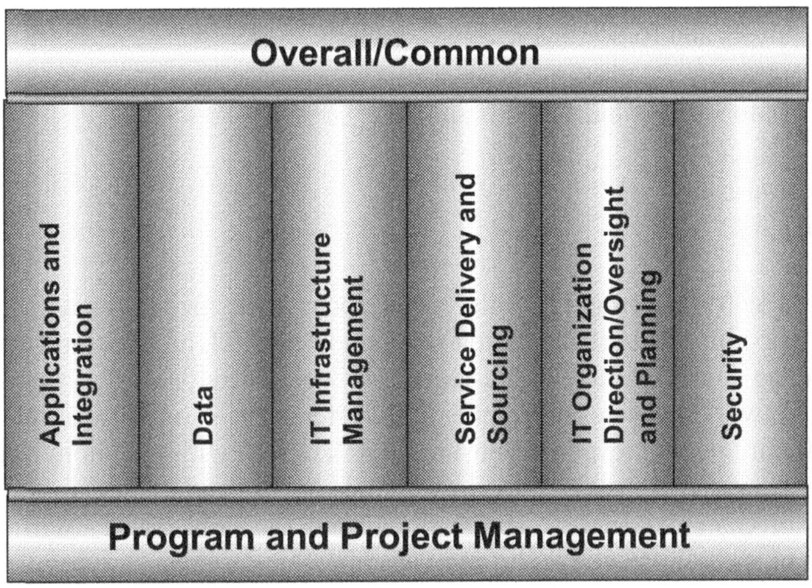


1. **Applications and Integration.** A number of IT applications are typical in a utility environment, including business and administrative systems, as well as systems to support operations and maintenance functions. Successful implementation of the same application (even the same vendor) varies from utility to utility. Continued emphasis on integration and interoperability of applications will be key to effective IT solutions. Most organizations have established basic application development methodologies and corresponding specific development tools. This focus area also deals with the integration of many, if not all, IT business applications and services with other applications, as well as with Internet (World Wide Web), Intranet, and browser technology.

   - Software Applications
   - Application Development and Enhancement
   - Application Upgrading and Decommissioning

- e-Business
- Application Specific Training and Support
- Integration (including tools and procedures)

2. **Data.** Data is the raw material produced by one or more business processes that create and maintain it. It is both an output from business processes, as well as an input into business processes. Increasing data complexity and ambiguity, development of security policies, and increased need for integration between applications require increased rigor around data. Industry-wide agreement and standardization will continue to grow in importance due to increased use of supply-chain interactions and the cost advantage of uniform performance measurement/decision support systems.

    - Data Definition, Models, and Standards
    - Data Acquisition/Input
    - Data Business Rules
    - Data Storage
    - Data Analysis and Reporting
    - Data Integration
    - Data Transformation and Conversion
    - Data Management
    - Data Stewardship
    - Data Quality

3. **IT Infrastructure Management.** This broad focus area includes several elements and technologies: telecom, networking, wireless technologies, IT architecture, and security. It includes managing the IT infrastructure assets, as well as automated systems for real-time monitoring of network, server, and application systems. It also includes the integration of traditional and more newly developed business applications with telephony (voice telephone communications) systems. Common examples include the integration of database information with customer service applications, and the concept of "unified messaging" where voice-mail, e-mail, fax and audio-mail are integrated with each other. The networking element incorporates network media (cabling), hardware (switches, routers, firewalls, etc.), network management systems, and other components and subsystems required to support the IT environment. The wireless component includes the use of wireless (radio-based, infrared-based, laser-based) technology to augment wired network topologies. It is most commonly applied toward providing mobile access to Internet-based, and enterprise-based IT applications and services. The IT architecture encompasses different approaches such as thin client-server, N-tier, and web services. Security is critical to all of these elements, especially as the Internet becomes more heavily relied upon for enterprise business applications and networking. This IT element deals with protection of data, systems, and trusted relationships.

    - IT Asset Tracking
    - Network Management
    - Telephony (IVR, ACD, Unified Messaging)

- Networks (Cabling, Routers, Firewalls)
- Wireless (Platforms and Integration)
- IT Architecture
- IT Service Desk (Help Desk, Desktop Management and Upgrades)

4. **Service Delivery and Sourcing.** Cost/benefit and risk assessment matrices have been developed to determine the tradeoffs in selecting service delivery and sourcing options. Service level and management level agreements are a critical aspect of this area in establishing requirements and the parameters around measuring the delivery of those requirements. This focus area also includes ongoing life cycle management of systems, whatever the delivery mechanism. Maintenance (both preventive and reactive) requires a systematic inventory of the environment (hardware, software, configurations, protocols) and continuous monitoring with an emphasis on identifying performance issues through "what if" scenarios associated with problem resolution, planned expansions, and application roll-out.

   - Procurement Policies and Procedures
   - Service Delivery Alternatives
   - Service Level Management
   - Systems Development Life-Cycle and Management

5. **IT Organization, Direction, Oversight and Planning.** Clarity of organizational goals and policies is essential for providing effective IT services on an ongoing basis. This clarity must also address the roles and goals of multiple jurisdictions. Work practices must support overriding policy principles. Technology policies are derived from business objectives and are used to lower the cost of technology ownership through the decision-making processes. Strategic planning services address all aspects of information technology, including current situation assessments, communications technology vision development, strategic requirements development and recommendations. This includes the hard technical areas of business systems as well as business planning and staffing, organizational issues, and budget development to address change management. Planning efforts move from strategic to tactical to operational to budgetary.

   - Governance and Policy Definition
   - Strategic/Organizational Plans
   - IT Vision, Mission, Goals
   - Budgeting (Capital, Operations and Annual)
   - Total Cost of Ownership
   - Business Continuity/Disaster Recovery
   - Roles and Responsibilities
   - Business Process Redesign

6. **IT Security:** Information technology security involves the protection of computer technology assets, including hardware, software, and data. IT security addresses the concepts of "CIA": systems and data **C**onfidentiality, **I**ntegrity, and **A**vailability. This involves making sure that only authorized users have access to systems and data; detecting and

preventing unauthorized access and/or tampering with data (including virus detection, prevention and recovery); data encryption; hardening systems; and disaster recovery.

Hardening systems involves actions such as changing hardware and software factory default settings, and generally customizing default levels of security. From a cyber-security perspective, disaster recovery includes processes involved with backing up data, performing data restores, and rebuilding systems from backups.

- Policies and Procedures
- Tools

7. **Program and Project Management:** Programs are large projects with multiple subprojects. Projects are well-defined team efforts to achieve a goal that has a distinct beginning and end, milestones and budget. Projects require well-defined roles and responsibilities and methodologies establishing the activities throughout the entire life cycle. A key component of effective programs and projects is a quality program, including testing and validation.

   - Program Definition and Management
   - Alignment with Strategic and Tactical Plan
   - Project Planning, Management and Tracking
   - Quality Control/Quality Assurance
   - Testing Strategies

8. **Overall/Common.** This focus area addresses overall and common IT issues. This includes organization structures and staffing approaches and performance management/measurement. In addition, most organizations have common methods and tools (for example, organizational communications and customer feedback mechanisms) that may be applied to any of the subsequent focus areas.

   - Hiring
   - Performance Management
   - IT Staff Development and Training
   - Organizational Structure and Reporting
   - Customer Satisfaction
   - Organizational Communications
   - IT Organizational Performance and Continuous Improvement
   - Technical Research
   - Marketing

# APPENDIX F
# PROJECT PARTICIPANT INFORMATION TECHNOLOGY ACCOMPLISHMENTS AND CHALLENGES

All participating utilities were surveyed regarding their most significant or recent information technology accomplishments, as well as the significant information technology challenges they face. Accomplishments fell into the following focus areas: applications and integration; IT infrastructure management; service delivery and sourcing; IT organizational direction/oversight and planning; and security. Obstacles fell into the following focus areas: overall/common; program and project management; applications and integration; service delivery and sourcing; IT organization/direction/oversight and planning, and IT security. A summary of their responses is provided below.

## ACCOMPLISHMENTS: APPLICATIONS AND INTEGRATION

### Work Management/GIS

- Selected a CMMS/FIS application
- Implementing our CMMS using process control systems technology standards
- Selected and now implementing a Work Order/Inventory Management system. This included performing an extensive "needs review."
- Modernized GIS to support future CMMS/CIS implementations. Efforts included data gathering and scrubbing; geospatial data modeling, upgrading to the latest Arc/Info version for our rollout, and a pilot of GPS asset location
- Implemented mini-GIS and hydraulic model within engineering (not IT)
- Implementing a City-wide GIS
- Integrated our mapped data with scanned as-built and engineering drawings from the EDMS effort

### Customer Service

- Expansion of our service area resulted in acquiring an additional billing system and more than doubling the number of customers serviced – we converted this system into the billing system we have historically used
- Designed, developed, and implemented a web-enabled CRM for 100,000 account base
- Justified and currently implementing a 70,000 point fully automated radio frequency meter reading system which leverages cellular phone technologies with radio frequency based fixed network technology
- Implemented customer e-Billing
- Implemented electronic billing (electronic bill presentment and payment) and EZ Pay (provides auto-deduct from customer's checking or savings)
- Interfaced IVR to financial system to provide customer's automated account information 24/7
- Implemented new call center software

- Implemented on-line water payment receipting system

**Other Business Applications**

- Acquiring new financial accounting system
- Implementing time and attendance system to automate paper time sheet procedures.
- Integrated time and attendance with new Financial Accounting system
- Implemented new budget system
- Selected enterprise-wide document management software – now implementing it in one Department
- Developed Environmental Monitoring for Public Access and Community Tracking (EMPACT) website
- Implemented help desk

**Accomplishments: IT Infrastructure**

- Developed new IT architecture to support separation of the retail and wires portions – driven by electrical deregulation
- Initiated, coordinated, and implemented a multi-department collaborative project to design and install WAN for four Departments
- Implemented and upgraded IT infrastructure including servers, network, operating system and supporting software
- Upgraded computerized access control and security system
- Upgraded phone system
- Standardizing on a single operating environment
- Upgraded mainframe computer ensuring >5 years growth
- Installed redundant fiber optics communication system to water plant

**Accomplishments: Service Delivery and Sourcing**

- Published SLAs with verifiable metrics and performance criteria

**Accomplishments: IT Organization Direction/Oversight and Planning**

- Established department's Information Technology Steering committee (cross-functional group of executives and managers) and Technology Advisory Group (cross-functional group of business and IT professionals). Now working to establish their mission.
- In second year of five year IT strategic plan with fifteen threads
- Conducted extensive review of control system
- Developed automation master plan
- Developed standards for process control systems technology and are now applying it to other applications
- Reorganized our IT Services section

**Accomplishments: IT Security**

- Completed information security related vulnerability assessments and implemented the majority of the recommendations

**Accomplishments: Sample Results:**

- Reduced operating personnel by 20 percent
- Realized a significant reduction in labor, chemical and energy costs due to plant automation enabled by an integrated IT system
- Improved customer service, response time and uptime

## BIGGEST IT OBSTACLES TO OVERCOME OR OPPORTUNITIES TO EXPLOIT

**Obstacles: Applications and Integration**

- Integrating business and process control applications to provide a seamless, stable computing environment
- Acquiring and implementing a browser interface tool for better access and management reporting of data across applications and a data warehouse
- Developing a wireless field operation system
- Implementing a wireless field operation system with existing resources
- Implementing an asset management system
- Integrating GIS into utility's business as a tool for all spatial data requirements
- Overall improvements in communication systems
- Developing Intranet and Internet web site(s) to support internal/external communications
- Using the Internet for e-commerce opportunities
- EBPP
- Providing customer account access over Internet

**Obstacles: Service Delivery and Sourcing**

- Increasing service demands in an environment where budgets remain flat
- Meeting increasing service demands from customers in an environment where budgets remain flat
- Better understanding of IT operating and maintenance requirements
- Providing excellent customer service to customers

**Obstacles: IT Organization Direction/Oversight and Planning**

- Creating and maintaining a consistent IT vision

- Working within a larger enterprise that has different IT priorities than the utility's IT priorities, and one that determines how much the utility will budget and spend for technology based on the resource limitations it has placed on the general fund departments
- Impact of varied priorities on IT budget
- Obtaining user's buy-in to cost benefits
- Having an effective cost justification or ROI process for IT projects
- Improving top management support through better understanding of IT operating and maintenance requirements and need for replacements and upgrades in five year intervals (as opposed to 20 to 75 year intervals on traditional operating equipment)
- Managing changes in workflow, practices, and people - including both IT and users.
- Dealing with the change management required as new tools are implemented. It is more than just getting the user needs (which are often based on limitations of current technologies)
- Leveraging the newer technologies and handling the people change issues to balance costs to IT and costs to the business
- Addressing the rapidly growing demand from other units in our organization who want or require new technology. The list of active projects continues to grow exponentially straining our ability to train project managers, mentor staff, and effect knowledge transfer.

**Obstacles: IT Security**

- Assessing and ensuring system security
- Dealing with the increasingly complex water security environment
- Increasing viruses and hacking activities in the Internet-connected world

**Obstacles: Program and Project Management**

- IT staffed trained in project management
- Ad-hoc allocating of resources to unplanned non-IT projects
- More effective and accurate IT project estimating including appropriate consideration for the cost of managing risks
- Developing and implementing a project post-implementation review process – verifying that benefits are achieved; rolling lessons learned into project estimating process
- Effective project management to handle increased IT project requests/backlog
- Acquiring and implementing Program Management software and methodologies to manage $4.5 Billion worth of infrastructure enhancement and replacements

**Obstacles: Overall/Common**

- Having enough IT resources to accomplish goals
- Lack of trained staff and inability to offer training for present staff

- Developing staff that can quickly and appropriately respond to changing technologies
- Keeping staff skills in line with project and user needs
- Upgrading skills; providing training to support migration from legacy systems to web based applications
- Cross-training to support knowledge transfer
- Digesting a major enterprise reorganization that resulted in centralization of IT services; realizing the benefits of this reorganization
- Properly marketing new systems (revenue generation) considering strategic corporate direction and existing regulations

# APPENDIX G
# EXAMPLE BUSINESS CASE WORKSHEETS

Example answers to the business case worksheet are provided below. They are taken from examples the participating utilities developed in an initial workshop. They address three focus areas/elements:

- Overall/Common: Customer Satisfaction
- Applications and Integration: Application, Development and Enhancement
- Data: Data Integration

Each of these worksheets is a starting point, that will ultimately lead to a project plan being developed.

## FOCUS AREA A – OVERALL/COMMON: CUSTOMER SATISFACTION

### Why is Improvement Needed?

- Unclear or inaccurate customer views and needs regarding IT
- Cost savings goals
- Need better communications between IT and others
- Better external customer satisfaction

### What will be the results/benefits?

- Increased communication
- Increased confidence in IT staff
- Workforce will have greater satisfaction in doing their job – improving their skills
- Reduced costs
- Clearer priorities

### How will improvements be achieved?

- Conduct surveys to identify actions
- Acquire software to help track performance
- Identify and implement a champion to improve customer satisfaction
- Set targets for performance improvements; measure them; analyze and post results and resulting actions
- Celebrate wins
- Report back to customers – know their needs
- Develop service level agreements with customers

**What are necessary resources and timeframe?**

- Some type of help desk software
- Analysis of data that is gathered
- Contract assistance (to help get started in setting up process)
- Intranet site
- Six months to one year

**How will success be measured?**

- Survey results will go up
- Rate of response to surveys will go up
- Increased external customer satisfaction
- Number of hits on the internet site will increase
- Decreased response time on the part of help desk
- Percent of action plans that are carried out will increase
- Monthly reports on how the staff is doing will reflect results

## FOCUS AREA B – APPLICATIONS AND INTEGRATION: APPLICATION, DEVELOPMENT AND ENHANCEMENT

**Why is Improvement Needed?**

- IT is leading the change process, but nobody else is really involved (everyone else is relying on IT to solve business unit problems)
- Tools, standards and guidelines do not uniformly exist or are not followed
- We need to institutionalize common practices– we are ad-hoc and not systematic in our approaches
- Process for inclusion of users/management is not consistent or frequent enough

**What will be the results/benefits?**

- More informed managers of the business units
- Clear direction, roadmap or structure
- Less effort and time getting the stakeholders all on the same page
- Building the structure within IT to deliver the solution will also help business units identify their structures
- Process will help IT become more service oriented (end user benefits)
- Users will be allowed to participate in application development process

**How will improvements be achieved?**

- Establish a common application selection, upgrade and implementation methodology
- Staff, time and baseball bat – need force!
- Requires executive sponsorship – buy in from top

- Must be priority for everyone (all levels)

**What are necessary resources and timeframe?**

- Establish schedule for the process and manage it
- Establish a committee mandated by executive from above

**How will success be measured?**

- Alignment with management, business units and IT programs (questionnaire)
- No surprises within the ranks (questionnaire)
- Fewer disruptions of daily operations (questionnaire)
- Improved accountability for results (new reports)
- Hidden project costs will be identified – better project estimates (budget, schedule and resources)
- Number of project plans following the common methodology
- Project is on schedule and under budget

## FOCUS AREA C - DATA: DATA INTEGRATION

**Why is Improvement Needed?**

- We have silo'ed applications, so we build custom interfaces to each application; lose data relationships and waste time re-inventing the wheel
- Have large amount of inconsistent or incorrect data
- We don't gain the business value from having data integrated
  - Asset management is separate from maintenance management – we have to re-enter data, update it multiple times, and reconcile data discrepancies
  - We need to "Harvest" our infrastructure – we've got a great infrastructure but don't reap the benefits

**What will be the results/benefits?**

- Reduced Cost (ROI to business) and (from an IT cost savings) as a result of quality data and integration
- Improved quality of data – people will trust the system more and spend less time keeping their own independent records
- Less time reconciling inconsistent systems

**How will improvements be achieved?**

- Executive sponsorship
- "Bull Dog" – the project will be driven to completion (all aspects of the project)
- Tenacity Master
- Communication

- Develop and implement an integration strategy and roadmap that includes business users and technology staff; pilot it first

**What are necessary resources and timeframe?**

- Subject matter experts and management to balance competing priorities
- Key business unit representatives and data stewards
- Executive/management support
- Top IT resources (internal or external)
- Five year timeframe

**How will success be measured?**

- Decision support across business units
- Fewer equipment breakdowns because of age – proactively replacing prior to breakdown, for example, replacing pumps before they break down
- Data quality and data use will improve because they are supported by business function
- Enterprise data map will exist, defining the data use

## REFERENCES

APQC (American Productivity & Quality Center), "The Benchmarking Code of Conduct". Houston, TX http://www.orau.gov/pbm/pbmhandbook/apqc.pdf

Brueck, T., Riddle, W., Paralez, L. American Water Works Association Research Foundation (AwwaRF), 2002. *Consortium Benchmarking Methodology Guide*, Denver, CO.

SEI (The Software Engineering Institute), The Capability Maturity Model® for Software (CMM or SW-CMM ). Software Engineering Institute, Carnegie Mellon University, Pittsburgh, PA 15213-3890 http://www.sei.cmu.edu/cmm/cmms/transition.html

Project Management Institute (PMI), Project Management Body of Knowledge (PMBOK) manual, 2000 Edition. http://www.pmi.org/prod/groups/public/documents/info/pp_pmbokguide2000excerpts.pdf

Infrastructure Technology Information Library (ITIL), June 27, 1995 http://www.instantweb.com/foldoc/foldoc.cgi?Information+Technology+Infrastructure+Library

### GENERAL INFORMATION TECHNOLOGY SOURCES

Association for Information Systems **http://www.aisnet.org**

Professional organization whose purpose is to serve academics specializing in Information Systems

Carnegie Mellon Software Engineering Institute: **www.sei.cmu.edu**;

Mission: To advance the practice of software engineering and to make the acquisition, development, and sustainment of software-intensive systems *predictably better, faster, and cheaper* for the DoD.

Center for Information Technology **www.cit.nih.gov/**

Mission: To provide, coordinate, and manage information technology, and to advance computational science

Center for Technology in Government **http://www.ctg.albany.edu/**

CTG is an applied research center devoted to improving government and public services through policy, management and technology innovation.

Chief Information Officers Council, **www.cio.gov**

Established by Executive Order 13011 (1996), the CIO Council serves as the interagency forum for improving practices in the design, modernization, use, sharing, and performance of Federal Government agency information resources. Their role includes developing recommendations for information technology management policies, procedures, and standards; identifying opportunities to share information resources; and assessing and addressing the needs of the Federal Government's IT workforce.

CIO Magazine: **www.cio.com**

Resource compilation for information management and technology

Computerworld Magazine: **www.computerworld.com/**

An information services company targeting IT professionals

Data Warehousing Institute: **www.dw-institute.com**

Provides education and training in the data warehousing and business intelligence industry including strategies, techniques, and tools required to design, build, and maintain data warehouses.

Federal Computer Week: **www.fcw.com/**

FCW Government Technology Group publishes two magazines Federal Computer Week and Government E-Business for federal, state and local governments and information technology companies that do business with the public sector.

Forrester Research, Inc.: **www.forrester.com**

An independent research firm that analyzes the future of technology change and its impact on businesses, consumers, and society.

Gartner Group Interactive: **www.gartnerweb.com**

Identifies and analyzes trends and technologies that have shaped the course of business

General Accounting Office **www.gao.gov/**

Agency that works for Congress and the American people. Congress asks GAO to study the programs and expenditures of the federal government. GAO is independent and nonpartisan.

Government Computer News **http://www.gcn.com/**

Online journal

Information Technology Association of America: **www.itaa.org**

Trade association representing the U.S. IT industry; web site provides information about the IT industry, its issues, association programs, publications, meetings, and seminars.

Information Technology Resources Board **http://www.irtb.gov**

Group of senior IT, acquisition, and program managers that provide at no cost to agencies, peer reviews of major Federal IT systems.

Info-tech Research Group: **http://www.infotechadvisor.com/index.cfm**

Advisory service focused on providing critical and relevant news for IT professionals.

Information Week **http://www.informationweek.com/**

Journal on business technology

Institute of Electrical and Electronic Engineers, **http://www.ieee.org/**

Authority in technical areas ranging from computer engineering, biomedical technology and telecommunications, to electric power, aerospace and consumer electronics, among others.

Institute of Internal Auditors **http://www.theiia.org/**

Association focused on internal auditing, governance and internal control, IT audit, education, and security.

Institute of Management and Administration (IOMA) **http://ioma.com/**

Publish a broad range of information products for business professionals designed to improve the efficiency and to enhance the financial performance of the organization.

IT Toolkit **http://www.ittoolkit.com/**

Tools, techniques, and resources related to IT. Free registration provides access to many IT how-to, best practice/lesson learned articles.

MIT Technology Review **http://www.techreview.com/**

Emerging technologies journal

National Association of State Chief Information Officers **https://www.nascio.org/**

NASCIO represents state chief information officers and information resource executives and managers from the 50 states, six U. S. territories, and the District of Columbia. State. Their mission is to shape national IT policy through collaborative partnerships, information sharing and knowledge transfer across jurisdictional and functional boundaries.

National Institute of Standards and Technology **http://www.nist.gov/**

Non-regulatory federal agency; their mission is to develop and promote measurements, standards, and technology to enhance productivity, facilitate trade, and improve the quality of life.

NIST Information Technology Laboratory **http://www.itl.nist.gov/**

ITL works with industry, research, and government organizations to make technology more usable, secure, scalable, and interoperable.

Office of Government-wide Policy, General Services Administration: http:// **www.gsa.gov**

Assist agencies better serve the public by offering, at best value, superior workplaces, expert solutions, acquisition services and management policies

Office of Information Technology, General Services Administration: **www.itpolicy.gsa.gov**

OIT promotes strategic management and effective use of Federal information technology through the collaborative development and execution of Government-wide programs and functions.

Service Level Management Learning Community **http://nextslm.org/**

Online learning community about Service Level Management.

Telecommunications Industry Association: **www.tiaonline.org**

Trade association in the communications and information technology industry

## IT POLICY, GOVERNANCE, AND PLANNING

GAO, "Maximizing the Success of Chief Information Officers: Learning From Leading Organizations," GAO-01-376G, February 2001.

Critical success factors, both organizational and operational, and leading principles for successful information management gleaned from private sector organizations.

IT Governance Institute, "Board Briefing on IT Governance," **http://www.itgovernance.org/boardbriefing.pdf**

Written for executive management to define why IT governance is important, what the issues are and what responsibilities management has

IT Governance Institute, "Information Security Governance: Guidance for Boards of Directors and Executive Management," **http://www.itgi.org/infosecurity.pdf**

Describes what managers may expect from their enterprise's information security program; how to implement an effective information security program, how to evaluate their own status with regard to a security program and how to decide what security program is desired.

IT Governance Portal **http://www.itgovernance.org/**

Premise: Effective enterprise governance focuses individual and group expertise and experience where it can be most productive... monitors and measures performance... provides assurance to critical issues. Information technology, long considered solely an enabler of an enterprise's strategy, must now be regarded as an integral part of that strategy. This web site offers information and resources showing how to efficiently and effectively deploy secure, reliable information and applied technology.

## METHODOLOGIES

COBIT (Control Objectives for Information and Related Technology) Framework
http://www.isaca.org/ct_proj.htm

The COBIT project is supervised by international representatives from industry, academia, government and the security and control profession. The project developed the COBIT Framework. The IT Governance Institute provides overall project guidance.

COBIT Steering Committee and IT Governance Institute, "Control Objectives," July 2000.
34 Control Objectives, one for each IT process grouped Premise: IT resources need to be managed by a set of naturally grouped processes. Framework includes a set of 34 Control Objectives, one for each IT processes grouped into planning and organization, acquisition and implementation, delivery and support, and monitoring.

COBIT Steering Committee and IT Governance Institute, "Framework," July 2000.
Premise: IT resources need to be managed by a set of naturally grouped processes. Framework includes a set of 34 Control Objectives, one for each IT processes grouped into planning and organization, acquisition and implementation, delivery and support, and monitoring.

COBIT Steering Committee and IT Governance Institute, "Implementation Tool Set," July 2000.
Lessons learned from organizations that applied COBIT in their work environments. Includes two tools – Management Awareness Diagnostic and IT control Diagnostic.

COBIT Steering Committee and IT Governance Institute, "Management Guidelines," July 2000.
Describes a framework for how to control IT so that it meets and delivers an organizations information needs while at the same time manage risks and secure the infrastructure

NASCIO, "Adaptive, Enterprise Architecture Development: Enterprise Architecture Development Tool Kit Version 1.01," **https://www.nascio.org/publications/index.cfm**, November 30, 2001.
Explores governance structures and provides examples of state and local government governance. Instruction set, templates and samples are provided.

## IT SERVICE DELIVERY AND SOURCING

GAO, "Leading Commercial Practices for Outsourcing of Services," GAO-02-214, November 2001. Identifies leading commercial practices for acquiring IT services.

Hiles, Andrew. The Complete Guide to IT Service Level Agreements, Connecticut: Rothstein Associates, 1999. Comprehensive guide to developing service level agreements.

LaBounty, Char, How to Establish and Maintain Service Level Agreements, Colorado: Help Desk Institute, 1999. Step-by-step process for developing service level agreements.

NASCIO, "Buying Smart: State Procurement Reform Saves Millions,"
**https://www.nascio.org/publications/Buying%20Smart%201.pdf**
Outlines recommendations to reform the government procurement process for information technology.

NASCIO, "Buying Smart: Blueprint for Action,"
**https://www.nascio.org/publications/Buying%20Smart%202.pdf**
Outlines innovative procurement strategies employed by state governments in the acquisition of information technology commodities.

Service Level Management, "Templates," **http://nextslm.org/templates.html**
Includes customer satisfaction survey, internal service level agreement, outline of a service level agreement, internal service level agreement (short).

IT Infrastructure Library, **http://www.itil-itsm-world.com/**
Provides a series of documents used to aid the implementation of a framework for IT Service Management (ITSM).

## IT BUSINESS PROCESS IMPROVEMENT

Carnegie Mellon Software Engineering Institute, "The Ideal Model: A Practical Guide for Improvement," **http://www.sei.cmu.edu/ideal/ideal.bridge.html**
Describes a life-cycle model for software process improvement based on the Capability Maturity Model.

Carnegie Mellon Software Engineering Institute, "IDEAL: A User's Guide for Software Process Improvement," **http://www.sei.cmu.edu/pub/documents/96.reports/pdf/hb001.96.pdf** , February 1996.
Describes the major activities of software process improvement

CIO Council, "Capital Planning and IT Investment Self Assessment," **http://cio.doe.gov/capitalplanning/TriFold/trifold.htm**
A tool for agencies to use in improving and maturing their internal capital planning and IT investment processes.

## IT PERFORMANCE MANAGEMENT

Center for Advanced Purchasing Studies, "Computer/Telecommunications Equipment/IT Services Industry Demographics," **http://www.capsresearch.org/BenchPDFs/Com_Tel_IT5.pdf** , 1999.

Center for Advanced Purchasing Studies, "Purchasing Performance Benchmarking Study for Municipal Governments," **http://www.capsresearch.org/BenchPDFs/MuniGov4.pdf** , 2000.

Center for Advanced Purchasing Studies, "Purchasing Performance Benchmarking Study for Utilities," **http://www.capsresearch.org/BenchPDFs/Utility2.pdf** , 2001.

Center for Strategic Management, "A Guide to Developing Effective IT Performance Measures," **http://irm.cit.nih.gov/itmra/itperfgu.doc**
This guide serves as a tool for designing and implementing IT performance measures. The underlying principles emphasize use of performance information for investment analysis and administrative decision-making.

GAO, "Executive Guide: Measuring Performance and Demonstrating Results of Information Technology Investments," GAO/AIMD-98-99, March 1998.

Setting performance targets, designing efficiency and effectiveness measures, systematically and accurately measuring outcomes, and using the results for informed decision-making in IT products, services, and delivery processes.

Gold, Robert S. "Enabling the Strategy-Focused IT Organization," **http://www.bscol.com/invoke.cfm?id=095FAE17-1322-11D5-A8D600508BDC96C1**, Balanced Scorecard Collaborative.
Using the Balanced Scorecard to correlate IT value to business strategy.

GSA, "Performance Based Management: Eight Steps to Develop and Use Information Technology Performance Measures Effectively," **http://www.gsa.gov/Portal/content/pubs_content.jsp?contentOID=116820&contentType=1008&PMKZ=1**
Guide for understanding performance measurement and who develop and use performance measures for IT projects.

Procurement Executives Association, "Guide to a Balanced Scorecard Performance Management Methodology," **http://oamweb.osec.doc.gov/bsc/**.

Guide describes the business system assessment program, implementation procedures, evaluation standards, and reporting process and provides a conceptual framework for performance management.

Project Management Institute **http://www.pmi.org**

Project Management Institute (PMI) is the world's leading not-for-profit project management professional association, with over 95,000 members worldwide. PMI members are individuals practicing and studying project management in many different industry areas, including aerospace, automotive, business management, construction, engineering, financial services, information technology, pharmaceuticals and telecommunications.

## IT OPERATIONS AND MAINTENANCE

BMC Software, "Best Practices in Infrastructure Management," **http://www.cio.com/sponsors/080801_bmc.html** or **http://www.bmc.com/assessment/im/index.cfm**

Objective: to develop a baseline of infrastructure management practices information to develop an online assessment tool. Survey focused on: networks, platforms or servers, storage, databases, middleware and IT security, and integration of management approaches. Data was collected on the effectiveness of monitoring managing, optimizing and making predictions regarding elements of the technology infrastructure.

GAO, "Improving Mission Performance Through Strategic Information Management and Technology: Learning From Leading Organizations," GAO/AIMD-94-115, May 1994.

Lessons learned from private or public organization on how to apply information technology to improve mission performance.

GAO, "Information Technology: Best Practices Can Improve performance and Produce Results," GAO/T-AIMD-96-46, February 1996.

How best practices can be effectively used to improve the management of information technology.

ITRB, "Practical Strategies for Managing Information Systems," **http://www.itrb.gov/documents/itrbps%2Ehtm**.

Lessons learned for effective delivery of mission critical information systems.

ITRB, "Managing Information Systems: A Practical Assessment Tool," **http://www.itrb.gov/documents/feb10ast%2Edoc** , February 1999.

Evaluative framework and checklist for assessing information systems' strategy, leadership and technology.

NASIRE, "Best Practices in the Use of Information Technology in State Government," **https://www.nascio.org/publications/bestpractices.cfm**

Best practices in the categories: accessibility, communications infrastructure, digital government, information architecture, innovative use of technology, IT professional retention and recruitment, public/private partnership, service applications, state planning and management initiatives.

## IT PROGRAM AND PROJECT MANAGEMENT

CSC, "14th Annual Survey of IS Management issues: 2001 Critical Issues," **http://www.csc.com/techleader/14.shtml**
Examines the IS organization, information security, key IT initiatives and provides industry perspectives.

CIO Council Capital Planning and IT Investment Committee, "Implementing Best Practices: Strategies at Work," June 1998.
Describes best practices in capital planning projects of 6 federal agencies

Center for Technology in Government, "Making Smart IT Choices: A Handbook," **http://www.ctg.albany.edu/resources/smartit.pdf**
Handbook designed to help you work through the information analysis process, identify and evaluate technology that will meet the need in the most effective, cost-effective way.

GAO, "Assessing Risks and Returns: A Guide for Evaluating Federal Agencies' IT Investment Decision-making," GAO/AIMD-10.1.13 February 1997.
Provides a structure for evaluating and assessing how well an agency is selecting and managing its IT resources and to identify specific areas where improvements can be made.

GAO, "Information Technology Investment Management: An Overview of GAO's Assessment Framework," GAO/AIMD-00-155, May 2000.
Decision-making process to assure that IT funds are invested and managed to achieve high value outcomes at acceptable costs.

GAO, "Information Technology Investment Management: A Framework for Assessing and Improving Process Maturity," GAO/AIMD-10.1.23, May 2000.
Builds on GAO/AIMD-10.1.13. ITIM Decision-making process to assure that IT funds are invested and managed to achieve high value outcomes at acceptable costs.

IEEE, "Recommended Practice for Software Requirements Specifications," IEEE 830-1998.

IEEE, "Standard for Software Project Management Plans," IEEE 1058-1998.

ITRB, "Assessing the Risks of Commercial Off the Shelf Products," **http://www.itrb.gov/documents/cotsfinal%2Edoc** December 1999.
Checklist for evaluating COTS.

ITRB, "Project Management for Mission Critical Systems," **http://www.itrb.gov/documents/itrbpmpr%2Ehtm** , April 2001.
Provides a high-level framework for project management of information systems projects.

OIT, "An Analytical Framework for Capital Planning and Investment Control for Information Technology," **http://www.itpolicy.gsa.gov/mke/caplan1.htm**
Framework for capital planning with an emphasis on financial analysis.

## IT QUALITY CONTROL/ASSURANCE

Information Systems and Audit and Control Association "Standards for Information Systems Auditing, " **http://www.isaca.org/standard/enstandards.pdf**

Information Systems and Audit and Control Association, "IS Auditing Guideline – Corporate Governance of Information Systems, " **http://www.isaca.org/standard/guide1.pdf**

Information Systems and Audit and Control Association, "IS Auditing Guideline – Planning the IS Audit," **http://www.isaca.org/standard/guide2.pdf**

IEEE, "Standard for Software Verification and Validation," IEEE 1012-1998.

ISO/IEC, "Information Technology: Software Packages – Quality Requirements and Testing", ISO/IEC 12119-1998.

**ORGANIZATIONAL**

Computerworld, "15th Annual Salary Survey: More for the Money"
**http://www.computerworld.com/storyba/0,4125,NAV47_STO63423,00.html**

Computerworld, "2001 Job Satisfaction Survey,"
**http://www.computerworld.com/cwi/story/0,1199,NAV47_STO61742,00.html**

National Academy of Public Administration, "The Transforming Power of Information Technology: Making the Federal Government an Employer of Choice for IT Employees," **http://www.encmarketing.com/clients/napa/** , August 2001.
Explore options for attracting and retaining IT skills. – explores alternative pay systems and human resources management programs that allow federal government programs to compete.

**SECURITY**

GAO, "Executive Guide: Information Security Management – Learning From Leading Organizations," GAO/AIMD-98-68, May 1998.
Guide is designed to promote awareness of information security issues and to provide information to establish a management framework for more effective information security programs.

GAO, "Information Security Risk Assessment: Practices of leading Organizations," GAO/AIMD-00-33, November 1999.
Provides an overview of risk assessment process

NASCIO, "Information Security in State Government Information Technology,"
**https://www.nascio.org/publications/Buying%20Smart%202.pdf**, 1999
Includes state survey results about information security models; technologies to protect data in e-commerce environments; challenges in maintaining information security; policies to prevent improper release of government records.

NIST, CIT, "Guide for Developing Security Plans for Information Technology Systems," **http://csrc.nist.gov/publications/nistpubs/800-18/Planguide.PDF**, NIST 800-18, December 1998.

# ABBREVIATIONS

| | |
|---|---|
| ATM | Asynchronous transfer mode |
| BPR | Business process reengineering |
| BSA | Business services analysis |
| CAD | Computer aided design |
| CADD | Computer aided drafting and design |
| CASE | Computer-aided software engineering |
| CDM | Corporate data management |
| CDPD | Commercial cellular digital packet data |
| CEO | Chief executive officer |
| CIO | Chief information officer |
| CIS | Customer information systems |
| CMM | Capability maturity model |
| CMMS | Computerized maintenance management system |
| DBA | Database administrator |
| DBMS | Database management systems |
| DCS | Data control system, data collection system |
| DRP | Disaster recovery plan |
| DVD | Digital video disk |
| ERP | Enterprise resource planning |
| FAQ | Frequently asked questions |
| FTE | Full time employee |
| GB | Gigabyte |
| GHZ | Gigahertz |
| GIS | Geographic Information Systems |
| IS | Information system |
| IT | Information technology |
| ITIL | Infrastructure technology information library |
| ITSM | Information technology service management |
| LAN | Local area network |
| LIMS | Laboratory information management system |
| MG | Million gallons |
| MMS | Manufacturing messaging specification |
| NIST | National institute of standards and technology |
| OS | Operating system |
| PAN | Personal Area Network |
| PC | Personal computer |
| PDA | Personal Digital Assistant |
| PERT | Program evaluation and review technique |
| PMBOK | Project management body of knowledge |
| PMI | Project management institute |
| PMM | Project management manual |
| PMO | Project management office |
| PMP | Project management professional |
| RCA | Root cause analysis |
| RDBMS | Relational database management system |

| ROI | Return on investment |
|---|---|
| SAN | Storage area network |
| SCADA | Supervisory control and data acquisition |
| SDLC | Systems development life cycle |
| SEI | Software engineering institute |
| SLA | Service level agreement |
| SMS | Systems management server |
| SOP | Standard operating procedure |
| SPU | Seattle public utilities |
| SVDLC | Software validation development life cycle |
| SW | Software |
| TCO | Total cost of ownership |
| TCP/IP | Transmission control protocol/Internet protocol |
| TSLT | Technical services lateral team |
| VoIP | Voice over IP |
| VPN | Virtual private network |
| WAN | Wide area network |
| WBS | Work breakdown structure |
| WEP | Wired equivalent privacy |

IWA Publishing's authorised EU representative for General Product Safety Regulations is Diane D'Arras, 15 rue Duret, 75116 Paris, France, e-mail: safety@iwap.co.uk.

Printed and bound by CPI Group (UK) Ltd, Croydon, CR0 4YY
12/05/2026
02108893-0003